MONOGRAPHS ON STATISTICS AND APPLIED PROBABILITY

General Editors

D.R. Cox, V. Isham, N. Keiding, T. Louis, N. Reid, and H. Tong

1 Stochastic Population Models in Ecology and Epidemiology *M.S. Barlett* (1960)
2 Queues *D.R. Cox and W.L. Smith* (1961)
3 Monte Carlo Methods *J.M. Hammersley and D.C. Handscomb* (1964)
4 The Statistical Analysis of Series of Events *D.R. Cox and P.A.W. Lewis* (1966)
5 Population Genetics *W.J. Ewens* (1969)
6 Probability, Statistics and Time *M.S. Barlett* (1975)
7 Statistical Inference *S.D. Silvey* (1975)
8 The Analysis of Contingency Tables *B.S. Everitt* (1977)
9 Multivariate Analysis in Behavioural Research *A.E. Maxwell* (1977)
10 Stochastic Abundance Models *S. Engen* (1978)
11 Some Basic Theory for Statistical Inference *E.J.G. Pitman* (1979)
12 Point Processes *D.R. Cox and V. Isham* (1980)
13 Identification of Outliers *D.M. Hawkins* (1980)
14 Optimal Design *S.D. Silvey* (1980)
15 Finite Mixture Distributions *B.S. Everitt and D.J. Hand* (1981)
16 Classification *A.D. Gordon* (1981)
17 Distribution-free Statistical Methods, 2nd edition *J.S. Maritz* (1995)
18 Residuals and Influence in Regression *R.D. Cook and S. Weisberg* (1982)
19 Applications of Queueing Theory, 2nd edition *G.F. Newell* (1982)
20 Risk Theory, 3rd edition *R.E. Beard, T. Pentikäinen and E. Pesonen* (1984)
21 Analysis of Survival Data *D.R. Cox and D. Oakes* (1984)
22 An Introduction to Latent Variable Models *B.S. Everitt* (1984)
23 Bandit Problems *D.A. Berry and B. Fristedt* (1985)
24 Stochastic Modelling and Control *M.H.A. Davis and R. Vinter* (1985)
25 The Statistical Analysis of Composition Data *J. Aitchison* (1986)
26 Density Estimation for Statistics and Data Analysis *B.W. Silverman* (1986)
27 Regression Analysis with Applications *G.B. Wetherill* (1986)
28 Sequential Methods in Statistics, 3rd edition
G.B. Wetherill and K.D. Glazebrook (1986)
29 Tensor Methods in Statistics *P. McCullagh* (1987)
30 Transformation and Weighting in Regression
R.J. Carrol and D. Ruppert (1988)
31 Asymptotic Techniques of Use in Statistics
O.E. Bardorff-Nielsen and D.R. Cox (1989)
32 Analysis of Binary Data, 2nd edition *D.R. Cox and E.J. Snell* (1989)

33 Analysis of Infectious Disease Data *N.G. Becker* (1989)
34 Design and Analysis of Cross-Over Trials *B. Jones and M.G. Kenward* (1989)
35 Empirical Bayes Methods, 2nd edition *J.S. Maritz and T. Lwin* (1989)
36 Symmetric Multivariate and Related Distributions
K.-T Fang, S. Kotz and K.W. Ng (1990)
37 Generalized Linear Models, 2nd edition *P. McCullagh and J.A. Nelder* (1989)
38 Cyclic and Computer Generated Designs, 2nd edition
J.A. John and E.R. Williams (1995)
39 Analog Estimation Methods in Econometrics *C.F. Manski* (1988)
40 Subset Selection in Regression *A.J. Miller* (1990)
41 Analysis of Repeated Measures *M.J. Crowder and D.J. Hand* (1990)
42 Statistical Reasoning with Imprecise Probabilities *P. Walley* (1991)
43 Generalized Additive Models *T.J. Hastie and R.J. Tibshirani* (1990)
44 Inspection Errors for Attributes in Quality Control
N.L. Johnson, S. Kotz and X, Wu (1991)
45 The Analysis of Contingency Tables, 2nd edition *B.S. Everitt* (1992)
46 The Analysis of Quantal Response Data *B.J.T. Morgan*
47 Longitudinal Data with Serial Correlation—A state-space approach
R.H. Jones (1993)
48 Differential Geometry and Statistics *M.K. Murray and J.W. Rice* (1993)
49 Markov Models and Optimization *M.H.A. Davis* (1993)
50 Networks and Chaos—Statistical and Probabilistic Aspects
O.E. Barndorff-Nielsen, J.L. Jensen and W.S. Kendall (1993)
51 Number-theoretic Methods in Statistics *K.-T. Fang and Y. Wang* (1994)
52 Inference and Asymptotics *O.E. Barndorff-Nielsen and D.R. Cox* (1994)
53 Practical Risk Theory for Actuaries
C.D. Daykin, T. Pentikäinen and M. Pesonen (1994)
54 Biplots *J.C. Gower and D.J. Hand* (1996)
55 Predictive Inference—An introduction *S. Geisser* (1993)
56 Model-Free Curve Estimation *M.E. Tarter and M.D. Lock* (1993)
57 An Introduction to the Bootstrap *B. Efron and R.J. Tibshirani* (1993)
58 Nonparametric Regression and Generalized Linear Models
P.J. Green and B.W. Silverman (1994)
59 Multidimensional Scaling *T.F. Cox and M.A.A. Cox* (1994)
60 Kernel Smoothing *M.P. Wand and M.C. Jones* (1995)
61 Statistics for Long Memory Processes *J. Beran* (1995)
62 Nonlinear Models for Repeated Measurement Data
M. Davidian and D.M. Giltinan (1995)
63 Measurement Error in Nonlinear Models
R.J. Carroll, D. Rupert and L.A. Stefanski (1995)
64 Analyzing and Modeling Rank Data *J.J. Marden* (1995)
65 Time Series Models—In econometrics, finance and other fields
D.R. Cox, D.V. Hinkley and O.E. Barndorff-Nielsen (1996)

66 Local Polynomial Modeling and its Applications *J. Fan and I. Gijbels* (1996)

67 Multivariate Dependencies—Models, analysis and interpretation *D.R. Cox and N. Wermuth* (1996)

68 Statistical Inference—Based on the likelihood *A. Azzalini* (1996)

69 Bayes and Empirical Bayes Methods for Data Analysis *B.P. Carlin and T.A Louis* (1996)

70 Hidden Markov and Other Models for Discrete-valued Time Series *I.L. Macdonald and W. Zucchini* (1997)

71 Statistical Evidence—A likelihood paradigm *R. Royall* (1997)

72 Analysis of Incomplete Multivariate Data *J.L. Schafer* (1997)

73 Multivariate Models and Dependence Concepts *H. Joe* (1997)

74 Theory of Sample Surveys *M.E. Thompson* (1997)

75 Retrial Queues *G. Falin and J.G.C. Templeton* (1997)

76 Theory of Dispersion Models *B. Jørgensen* (1997)

77 Mixed Poisson Processes *J. Grandell* (1997)

78 Variance Components Estimation—Mixed models, methodologies and applications *P.S.R.S. Rao* (1997)

79 Bayesian Methods for Finite Population Sampling *G. Meeden and M. Ghosh* (1997)

80 Stochastic Geometry—Likelihood and computation *O.E. Barndorff-Nielsen, W.S. Kendall and M.N.M. van Lieshout* (1998)

81 Computer-Assisted Analysis of Mixtures and Applications—Meta-analysis, disease mapping and others *D. Böhning* (1999)

82 Classification, 2nd edition *A.D. Gordon* (1999)

83 Semimartingales and their Statistical Inference *B.L.S. Prakasa Rao* (1999)

84 Statistical Aspects of BSE and vCJD—Models for Epidemics *C.A. Donnelly and N.M. Ferguson* (1999)

85 Set-Indexed Martingales *G. Ivanoff and E. Merzbach* (2000)

Set-Indexed Martingales

GAIL IVANOFF
Professor of Mathematics and Statistics
University of Ottawa
Ontario, Canada

ELY MERZBACH
Professor of Mathematics and Computer Science
Bar-Ilan University
Ramat Gan, Israel

CHAPMAN & HALL/CRC

Boca Raton London New York Washington, D.C.

Library of Congress Cataloging-in-Publication Data

Ivanoff, B. Gail.
    Set-indexed martingales / B. Gail Ivanoff, Ely Merzbach.
       p. cm. — (Monographs on statistics and applied probability ; 85)
    Includes bibliographical references and index.
    ISBN 1-58488-082-1 (alk. paper)
    1. Martingales (Mathematics). I. Merzbach, Ely. II. Title.
III. Series.
QA274.5.I93 1999
519.2′87—dc21   99-44879
                                                                                                                                                           CIP

    This book contains information obtained from authentic and highly regarded sources. Reprinted material is quoted with permission, and sources are indicated. A wide variety of references are listed. Reasonable efforts have been made to publish reliable data and information, but the author and the publisher cannot assume responsibility for the validity of all materials or for the consequences of their use.

    Neither this book nor any part may be reproduced or transmitted in any form or by any means, electronic or mechanical, including photocopying, microfilming, and recording, or by any information storage or retrieval system, without prior permission in writing from the publisher.

    The consent of CRC Press LLC does not extend to copying for general distribution, for promotion, for creating new works, or for resale. Specific permission must be obtained in writing from CRC Press LLC for such copying.

    Direct all inquiries to CRC Press LLC, 2000 N.W. Corporate Blvd., Boca Raton, Florida 33431.

    Trademark Notice: Product or corporate names may be trademarks or registered trademarks, and are used only for identification and explanation, without intent to infringe.

© 2000 by Chapman & Hall/CRC

No claim to original U.S. Government works
International Standard Book Number 1-58488-082-1
Library of Congress Card Number 99-44879
Printed in the United States of America   1 2 3 4 5 6 7 8 9 0
Printed on acid-free paper

This book is dedicated to our spouses, Philip and Myriam, and to our children for their love, support and patience, without which this book would not have been possible.

Through wisdom a house is built.
Proverbs 24:3

Contents

Introduction		1
I General Theory		**7**
1 Generalities		**9**
1.1	Framework and Assumptions	9
1.2	Examples	15
1.3	The Hausdorff Metric	19
1.4	The Probability Structure	22
1.5	Stopping Sets	28
2 Predictability		**35**
2.1	A characterization by stochastic intervals	35
2.2	Announcability	40
2.3	Progressivity	43
2.4	Left-continuous processes	47
3 Martingales		**53**
3.1	Definitions	54
3.2	Classical properties	57
3.3	Stopping theorems	61
3.4	Examples	66
4 Decompositions and Quadratic Variation		**73**
4.1	Definitions	73
4.2	Admissible Functions and Measures	74
4.3	Compensator and Quadratic Variation	81
4.4	Discrete Approximations	86
4.5	Point Processes and Compensators	91
5 Martingale Characterizations		**97**

	5.1 Flows	98
	5.2 Brownian Motion	106
	5.3 The Poisson Process	107
6	**Generalizations of Martingales**	**113**
	6.1 Local Martingales	114
	6.2 Doob-Meyer Decompositions	116
	6.3 Quasimartingales	119
II	**Weak Convergence**	**131**
7	**Weak Convergence of Set-Indexed Processes**	**133**
	7.1 The Function Space $\mathcal{D}(\mathcal{A})$	134
	7.2 Weak Convergence on $\mathcal{D}(\mathcal{A})$	144
	7.3 Semi-Functional Convergence	147
8	**Limit Theorems for Point Processes**	**151**
	8.1 Strictly simple point processes	152
	8.2 Poisson limit theorem	160
	8.3 Empirical processes	169
9	**Martingale Central Limit Theorems**	**175**
	9.1 Central Limit Theorems	175
	9.2 The Weighted Empirical Process	181
References		**187**
Index		**207**

Acknowledgements

It is with great pleasure that we acknowledge the contribution of our doctoral students, Raluca Balan and Leah Burstein, in the preparation of this book. Thanks are also due to the second author's postdoctoral fellow, Dr. Diane Saada, for a careful reading of several chapters. Finally, we are very grateful to Dr. Dean Slonowsky, a former student of the first author, for a fruitful research collaboration and for permitting us to include many of his results in this book.

The first author thanks the National Sciences and Engineering Research Council of Canada for its financial support. The second author expresses his appreciation to the first author and to the University of Ottawa for their warm hospitality over the years, and to Bar-Ilan University for its support.

Introduction

The objective of this work is to take the initial steps in the development of a general theory of martingales indexed by a family of sets. Several diverse papers on this subject have been published, but very few with this 'general theory' point of view in mind. Here, in a unified setting, we will present results of about ten papers published by the authors (and co-authors) on this subject and we will include as well recent work by other researchers.

The theory of set-indexed martingales brings together two theories: the theory of set-indexed stochastic processes and the theory of martingales with a directed index set. These two theories were motivated by different problems; both were first developed in the fifties and sixties, and important results were obtained. The fact that set-indexed martingales possess the structure of both types of processes results in a rich and fruitful interaction. Here we are able to exploit the dynamical martingale structure of the processes together with the topological properties of the collection of index sets. Establishing a suitable framework was the key to the development of this more general structure.

The roots of the theory of set-indexed stochastic processes lie primarily in applications; they include: Gaussian processes (cf. Chentsov (1957), Yaglom (1957), Belyaev (1961), Dudley (1965), Rozanov (1967), Strait (1966), Nosko (1969), Wong (1969), Berman (1970), Cartier (1971), Kallianpur and Mandrekar (1974), Dobrushin and Surgailis (1979) and more recently works of Bass and Pyke (1984), (1985) and (1987), Alexander (1986), Adler (1990)); random fields (cf. Adler (1981), Rosen (1984)); random measures (cf. Kallenberg (1986), Blei (1989)); stochastic geometry (cf. Matheron (1975), Stoyan, Kendall and Mecke (1987)); spatial statistics and point processes (cf. Fisher (1972), Jagers (1974), Moran (1976), Papangelou (1976)). On the other hand, the principal motivation for the study of martingales indexed by a directed set was the theoretical interest of the subject (cf. Bochner (1955), Krickeberg

(1956), Helms (1958), Chow (1960), Licea (1969), Wichura (1969), Neuhaus (1971), Straf (1972), Gabriel (1977), Astbury (1978), Kurtz (1980a), Washburn and Willsky (1981), Norberg (1989a)).

The theory of two-parameter stochastic processes was a source of inspiration and a well-spring of examples for set-indexed martingales, since a process indexed by points of the positive quadrant of the plane can be viewed as a process indexed by rectangles with the origin as a corner of the rectangle. Therefore, an important part of this book consists of extending results of the two-parameter theory to our general framework. However, in our general setting, more elegant methods are proposed, and new and sometimes surprising results are obtained. The literature on the theory of two-parameter processes was very plentiful in the seventies and eighties: seminal articles were written by Wong and Zakai (1974) and Cairoli and Walsh (1975). The bibliography lists the work of later contributors in this field, including (in alphabetical order) Al Hussaini and Elliott, Allain, Bakry, Brossard, Cabana and Wschebor, Chevalier, Dalang, Dozzi, Ehm, Evstigneev, Follmer, Fouque, Frangos, Gikhman, Gundy, Gushchin, Guyon and Prum, Hajek, Hurzeler, Imkeller, Ivanoff, Kallianpur, Korezlioglu, Kurtz, Ledoux, Mandelbaum, Mandrekar, Mazziotto and Szpirglas, Merzbach, Meyer, Millet, Mishura, Neuhaus, Nualart, Orey, Park, Pitt, Pyke, Russo, Sanz, Smythe, Stoica, Sucheston, Sznitman, Tudor, Ustunel, Vanderbei, Vares, Yeh, Yor, Zbaganu and Zhuang, and Zhang. This is only part of a very extensive list.

Some difficult problems were solved for the two-parameter theory while others were not, but in any event the fundamental differences between the classical theory and the two-parameter theory were highlighted by the research done in the eighties, and for this reason alone the results are of interest.

Thus, the motivation for this work was two-fold: on the one hand, the theoretical aspects of the theory are rich and promising. On the other hand, set-indexed martingales have many applications, especially in the theory of spatial statistics and in stochastic geometry. It is clear that much work remains to be done and there are many important problems outstanding. In this monograph, we have tried to give our readers a foretaste of this subject, and it is our hope that it will motivate further development of the theory and applications of set-indexed martingales.

The book is essentially divided into two parts. Part I constitutes the 'general theory' and Part II presents results on weak convergence of sequences of martingales. The framework presented in Part I is kept as general as possible; for example, for most of Part I the martingale may be assumed to be indexed by subsets of a general σ-compact topological space T, whereas for the study of weak convergence in Part II, it must be assumed that T is metrizable. Part I is comprised of Chapters 1-6, and Part II of Chapters 7-9.

Part I begins with a chapter of generalities in which the framework, notation, and assumptions required for the general theory of martingales indexed by a family of sets are developed. The index set used throughout the book is a class $\mathcal{A}$ of compact connected subsets of T. We consider only a set-indexed analogue of the 'continuous'-parameter case (i.e. martingales indexed by $\mathbf{R}_+$); an analogue of discrete (i.e. $\mathbf{N}$-indexed) martingale theory would require a separate (albeit more straightfoward) development. The choice of the collection $\mathcal{A}$ is critical: its structure must be appropriate for a theory of martingales. On the one hand, in most instances it must be sufficiently rich to generate at least the Borel sets of T, while on the other hand, we frequently require it to be small enough to ensure fundamental properties of specific stochastic processes, such as the existence of a continuous version of the set-indexed Brownian motion. After examining the general framework, we describe the probability structure and introduce the critical concept of 'stopping'. As is often the case in mathematics, the most difficult aspect of the development of the theory is finding the 'right' definitions. As a result, the appropriateness of the framework may only become evident after reading several chapters of the book.

In Chapter 2, the fundamental concept of predictability is studied, and two different predictable σ-algebras are given which depend on the definition of the 'history' of a set. One of the predictable σ- algebras may be completely characterized via stochastic intervals. It is interesting to notice that for this important characterization, the topological structure of T is not really necessary. A section is devoted to the notion of announcability. In contrast to the classical one-parameter case, in our general context predictability and announcability are not equivalent. Two supplementary sections on progressivity and left-continuous processes are of theoretical interest, but are not required for the sequel.

Martingales are defined and studied in Chapter 3. Roughly speaking, in the set-indexed framework, there are three different types of martingales: strong martingales, martingales and weak martingales. All are essential for the theory. Strong and weak martingales are characterized by the existence of a measure on one of the predictable product spaces. According to the type of history used in the definition, we get either a strong or a weak martingale; most of their properties are connected with measure theory. On the contrary, martingales are defined using the partial-order of set-inclusion. No measure is involved and therefore other analytical properties are obtained. We begin this chapter by briefly reviewing some classical properties of martingales which have a straightforward extension to our framework. Next, a section is devoted to stopping theorems. Clearly, we get different results corresponding to the different types of martingales and now the rôle of stopping sets becomes clear. Finally we give some applications and examples. In particular, we emphasize two classical and fundamental processes: the set-indexed Brownian motion and the set-indexed Poisson process. These processes are well-known and have been extensively studied by many authors (see the bibliography). However, in our context, the martingale properties of these processes are developed, illustrating the richness of the theory.

In Chapter 4, extensions of the Doob-Meyer decomposition are studied. In the first section we introduce the compensator of a submartingale, and the quadratic variation of a square-integrable martingale. It will be seen that although the two concepts are closely related, unlike the one-dimensional case, the quadratic variation is not the compensator of a submartingale, and so must be considered separately. In the second section we introduce admissible functions. These functions constitute the key to the Doob-Meyer decomposition and the quadratic variation process, and so the main problem is to find conditions on the process which ensure that the finitely-additive admissible function can be extended to a measure. Next we prove the Doob-Meyer decompositions: under some assumptions, any strong submartingale can be decomposed uniquely into a sum of a martingale and a predictable increasing process, and similarly, a submartingale can be decomposed uniquely into a sum of a weak martingale and a predictable increasing process. Likewise, we show the existence of a quadratic variation process for both martingales and strong martingales. Two different proofs are presented: one involves proving that there exists a unique predictable increasing process which corresponds to the admissible function, and the second makes use of a discrete weak L^1-approximation. Finally, examples are considered illustrating both the Doob-Meyer decomposition and the quadratic variation process.

INTRODUCTION 5

Chapter 5 is devoted to martingale characterizations of the set-indexed Brownian motion and the Poisson process, which extend the corresponding characterizations of these processes on $\mathbf{R}_+$ due respectively to Lévy and Watanabe. The main tool used is a 'flow' which permits us to apply classical one-dimensional martingale techniques in the set-indexed framework. Flows reappear in the weak convergence results presented in Part II of the book.

The last chapter of Part I, Chapter 6, includes different generalizations of martingales. We begin with local martingales. In the two-parameter case, there are several non-equivalent definitions of local martingales, and we suggest one possible approach to localization for set-indexed processes. We are then able to prove a Doob-Meyer decomposition for a local submartingale. Finally, the various concepts of set-indexed quasimartingales are defined and studied in the last section of this chapter.

Part II opens with Chapter 7, which introduces two concepts of weak convergence. The first involves the function space $\mathcal{D}(\mathcal{A})$, which is a generalization of Skorokhod's function space $D[0,1]$, and consists of processes which are 'outer-continuous with inner limits'. Compactness criteria are developed, and sufficient conditions for functional convergence of a sequence of $\mathcal{D}(\mathcal{A})$-valued processes are given. However, it is noted that not all martingales are $\mathcal{D}(\mathcal{A})$-valued (not even the Brownian motion process!), and so for more general results a different approach must be taken, and a second mode of convergence, 'semi-functional' convergence, is introduced. Its definition is linked closely to the concept of flows.

In Chapter 8, we consider functional convergence of a sequence of point processes. We introduce 'strictly simple' point processes and show that convergence in finite dimensional distribution of a sequence of point processes to a strictly simple limit is equivalent to convergence in $\mathcal{D}(\mathcal{A})$. As an example, we prove a functional Poisson convergence theorem. This theorem is then applied to set-indexed empirical processes.

The last chapter of the book, Chapter 9, deals with martingale central limit theorems. We prove both functional and semi-functional CLT's for sequences of strong martingales, and observe that the behaviour of the sequence of quadratic variation processes determines asymptotic normality, just as it does for sequences of martingales on $\mathbf{R}_+$. Finally, as an application, a CLT is proven for a set-indexed weighted empirical process.

This monograph has been the result of a research collaboration of more than a decade between the authors. We have enjoyed putting together the pieces of the puzzle which resulted from the many papers we have written together and with others.

PART I

General Theory

CHAPTER 1

Generalities

This chapter introduces the general framework, notation, and assumptions required for the general theory of martingales indexed by a family of sets as developed in the papers of the authors and co-authors (Dozzi, Lin, Schiopu-Kratina and Slonowsky). As well, we will present many fundamental technical results which will be required throughout subsequent chapters, which consequently may make the chapter appear somewhat formidable at first glance. However, proofs may be omitted on a first reading and referred to only as needed, and examples will be used throughout to illustrate the various technical details.

1.1 Framework and Assumptions

We begin with a discussion of the index set. Throughout the book, T will denote a non-void sigma-compact topological space such that $\emptyset$ and T are the only sets which are both closed and open. $\mathcal{O}$, $\mathcal{K}$, $\mathcal{B}$ denote respectively the classes of open, compact and Borel sets of T. In some parts of the book, and particularly in Part II where we deal with weak convergence, it will also be assumed that T is a complete, separable metric space with metric d, but except where indicated, this is not required for Part I. When this assumption is required, it will be explicitly stated. Occasionally other structures such as a partial order or an algebraic structure may be imposed on T, but these will be defined and explained as required.

All processes will be indexed by a class $\mathcal{A}$ of compact subsets of T, and for *any* class of sets $\mathcal{A}$, the class of finite unions of sets from $\mathcal{A}$ will be denoted by $\mathcal{A}(u)$. In order to obtain an interesting theory of set-indexed martingales, the choice of the collection $\mathcal{A}$ is critical: its structure must be appropriate for a theory of martingales, it must be sufficiently rich in most instances to generate $\mathcal{B}$, the Borel sets of T, while at the same time small enough to ensure fundamental properties of specific stochastic processes, such as the existence of a continuous version (in $\mathcal{A}$) of the set-indexed Brownian motion. These considerations lead to the following definition:

Definition 1.1.1 *A nonempty class $\mathcal{A}$ of compact, connected subsets of T is called an* indexing collection *if it satisfies the following:*

1. $\emptyset \in \mathcal{A}$, and $A^\circ \neq A$ if $A \neq \emptyset$ or T. In addition, there is an increasing sequence (B_n) of sets in $\mathcal{A}(u)$ such that $T = \cup_{n=1}^{\infty} B_n^\circ$. (Hence, for every $A \in \mathcal{A}$, there exists $n = n(A)$ such that $A \subseteq B_n$.)

2. $\mathcal{A}$ is closed under arbitrary intersections and if $A, B \in \mathcal{A}$ are nonempty, then $A \cap B$ is nonempty. If (A_i) is an increasing sequence in $\mathcal{A}$ and there exists n such that $A_i \subseteq B_n$ for every i, then $\overline{\cup_i A_i} \in \mathcal{A}$. (Such a sequence (A_i) is called bounded.)

3. The σ-algebra generated by $\mathcal{A}$, $\sigma(\mathcal{A}) = \mathcal{B}$, the collection of all Borel sets of T.

4. Separability from above
   There exists an increasing sequence of finite subclasses $\mathcal{A}_n = \{A_1^n, ..., A_{k_n}^n\}$ of $\mathcal{A}$ closed under intersections and satisfying $\emptyset$, $B_n \in \mathcal{A}_n(u)$ (B_n is defined in 1. above and $\mathcal{A}_n(u)$ is the class of unions of sets in $\mathcal{A}_n$), and a sequence of functions $g_n : \mathcal{A} \to \mathcal{A}_n(u) \cup \{T\}$ such that

   (a) g_n preserves arbitrary intersections and finite unions (i.e. $g_n(\cap_{A \in \mathcal{A}'} A) = \cap_{A \in \mathcal{A}'} g_n(A)$ for any $\mathcal{A}' \subseteq \mathcal{A}$, and if $\cup_{i=1}^k A_i = \cup_{j=1}^m A_j'$, then $\cup_{i=1}^k g_n(A_i) = \cup_{j=1}^m g_n(A_j')$),
   (b) for each $A \in \mathcal{A}$, $A \subseteq (g_n(A))^\circ$,
   (c) $g_n(A) \subseteq g_m(A)$ if $n \geq m$,
   (d) for each $A \in \mathcal{A}$, $A = \cap_n g_n(A)$,
   (e) if $A, A' \in \mathcal{A}$ then for every n, $g_n(A) \cap A' \in \mathcal{A}$, and if $A' \in \mathcal{A}_n$ then $g_n(A) \cap A' \in \mathcal{A}_n$.
   (f) $g_n(\emptyset) = \emptyset \ \forall n$.

(Note: '$\subset$' indicates strict inclusion and '$\overline{(\cdot)}$' and '$(\cdot)^\circ$' denote respectively the closure and the interior of a set.)

Comments:

1. At first reading, the condition of 'separability from above' may seem somewhat complicated, but in reality it is very natural and is trivially satisfied in all of the examples which will be presented shortly. The rôle of the g_n's is to permit us to approximate each set in $\mathcal{A}$ 'strictly from above'.

FRAMEWORK AND ASSUMPTIONS

2. Definition 1.1.1 implies that our space T cannot be discrete, and that $\mathcal{A}$ is at least a continuum. All the examples are of this nature. The discrete case would be much simpler to study, as the separability condition would not be needed.

3. Note that any collection of sets closed under intersections is a semilattice with respect to the partial order of set inclusion. In the examples we have in mind, the semilattices $\mathcal{A}_n$ may be chosen so that $g_n(A) = \cap_{B \in \mathcal{A}_n(u), A \subseteq B^\circ} B$.

4. Denote $\emptyset' = \cap_n \cap_{A \in \mathcal{A}_n, A \neq \emptyset} A = \cap_{A \in \mathcal{A}, A \neq \emptyset} A$. (The second equality is a result of separability.) We have that $\emptyset' \in \mathcal{A}$ and since the sets in $\mathcal{A}$ are compact, by the finite intersection property, $\emptyset' \neq \emptyset$. Without loss of generality, we shall assume that $\emptyset' \in \mathcal{A}_n \; \forall n$. The rôle of $\emptyset'$ is central to the development of the martingale theory, and is analogous to that of 0 for martingales defined on $\mathbf{R}_+$. *Unless otherwise stated, we shall always assume that $\emptyset' \in \mathcal{A}'$, whenever $\mathcal{A}'$ is a sub-semilattice of $\mathcal{A}$.*

5. Trivially, since g_n preserves finite unions, $A \subseteq B \Rightarrow g_n(A) \subseteq g_n(B)$. As well, g_n has a unique extension to $\mathcal{A}(u)$ (the class of finite unions of sets in $\mathcal{A}$) which is defined as follows: for $B \in \mathcal{A}(u)$, $g_n(B) = \cup_{A \in \mathcal{A}, A \subseteq B} g_n(A)$. It is easily verified that g_n preserves finite intersections and unions of sets in $\mathcal{A}(u)$, and that $B \subseteq (g_n(B))^\circ$. Thus, if $A \in \mathcal{A}$, $B \in \mathcal{A}(u)$ and $A \not\subseteq B$, then since A is connected, $g_n(B) \cap A \not\subseteq B$. A continuity property of the extension of g_n is given in the following lemma.

Lemma 1.1.2 *If $B \in \mathcal{A}(u)$, then $B = \cap_n g_n(B)$.*

Proof. Let $B = \cup_{i=1}^k A_i$ be any representation of B. Since g_n preserves finite unions, it is easily seen that $g_n(B) = \cup_{i=1}^k g_n(A_i)$. Thus, if $x \in \cap_n g_n(B)$, then there exists i, $1 \leq i \leq k$ such that $x \in g_n(A_i)$ infinitely often. Since $(g_n(A_i))_n$ is a decreasing sequence, $x \in g_n(A_i)$, $\forall n$. But $B = \cup_{i=1}^k A_i = \cup_{i=1}^k \cap_n g_n(A_i)$ and so $x \in B$. □

In general, $\mathcal{A}(u)$ is not closed under countable intersections, so we will occasionally require a larger class $\tilde{\mathcal{A}}(u)$, which is the class of countable intersections of sets in $\mathcal{A}(u)$: i.e. $B \in \tilde{\mathcal{A}}(u)$ if there exists a sequence (A_n) in $\mathcal{A}(u)$ such that $\cap_n A_n = B$.

If $B \in \tilde{\mathcal{A}}(u)$, we define $g_n(B)$ as follows:

$$g_n(B) = \bigcap_{B' \in \mathcal{A}(u), B \subseteq B'} g_n(B').$$

This definition leads to additional continuity properties for g_n:

Lemma 1.1.3 (a) If $B \in \tilde{\mathcal{A}}(u)$, then $B = \cap_n g_n(B)$.
(b) For any class $\mathcal{D} \subseteq \tilde{\mathcal{A}}(u)$, $\cap_{B \in \mathcal{D}} B \in \tilde{\mathcal{A}}(u)$ and $g_m(\cap_{B \in \mathcal{D}} B) = \cap_{B \in \mathcal{D}} g_m(B)$.

Proof. (a) Clearly, $B \subseteq \cap_n g_n(B)$. There exists a countable subclass $\mathcal{B}'$ of $\mathcal{A}(u)$ such that $B = \cap_{B' \in \mathcal{B}'} B'$. Lemma 1.1.2 implies that $B' = \cap_n g_n(B')$ if $B' \in \mathcal{B}'$, so $B = \cap_{\mathcal{B}'} \cap_n g_n(B') = \cap_n \cap_{\mathcal{B}'} g_n(B')$. But by definition, $g_n(B) \subseteq g_n(B')$, $\forall\, B' \in \mathcal{B}'$, so $g_n(B) \subseteq \cap_{\mathcal{B}'} g_n(B')$. Finally, we have $\cap_n g_n(B) \subseteq \cap_n \cap_{\mathcal{B}'} g_n(B') = B$.
(b) First, for any $\mathcal{D} \subseteq \tilde{\mathcal{A}}(u)$,
$$\cap_{B \in \mathcal{D}} B = \cap_{B \in \mathcal{D}} \cap_n g_n(B) = \cap_n \cap_{B \in \mathcal{D}} g_n(B).$$
Now, since $\mathcal{A}_n(u)$ is a finite lattice, $\cap_{B \in \mathcal{D}} g_n(B) = B_n \in \mathcal{A}_n(u)$ and so $\cap_{B \in \mathcal{D}} B = \cap_n B_n \in \tilde{\mathcal{A}}(u)$.

Next, we begin by proving that if $B_n \in \mathcal{A}(u)\ \forall n$, then
$$\cap_n g_m(B_n) = g_m(\cap_n B_n), \quad \forall m. \tag{1.1}$$
By monotonicity it is clear that $g_m(\cap_n B_n) \subseteq \cap_n g_m(B_n)$. Conversely, by definition, $\cap_n g_m(B_n) = \cap_n \cup_{A \in \mathcal{A}, A \subseteq B_n} g_m(A)$. Thus, if $x \in \cap_n g_m(B_n)$, there exists a sequence (A_n) of sets in $\mathcal{A}$ such that $x \in g_m(A_n)$ and $A_n \subseteq B_n$, $\forall n$. Therefore, since g_m preserves countable intersections of sets in $\mathcal{A}$, $x \in \cap_n g_m(A_n) = g_m(\cap_n A_n) \subseteq g_m(\cap_n B_n)$, as required.

Now to complete the proof of (b), we remark that, as noted above, $\cap_{B \in \mathcal{D}} g_n(B) = B_n \in \mathcal{A}_n(u)$ and so $\cap_{B \in \mathcal{D}} B = \cap_n B_n$. Therefore,

$$\begin{aligned}
g_m(\cap_{B \in \mathcal{D}} B) &= g_m(\cap_n B_n) \\
&= \cap_n g_m(B_n) \text{ by (1.1)} \\
&= \cap_n g_m(\cap_{B \in \mathcal{D}} g_n(B)) \text{ by definition} \\
&= \cap_n \cap_{B \in \mathcal{D}} g_m(g_n(B)) \\
&\quad \text{since } \cap_{B \in \mathcal{D}} g_n(B) \text{ is a finite intersection} \\
&= \cap_{B \in \mathcal{D}} \cap_n g_m(g_n(B)) \\
&= \cap_{B \in \mathcal{D}} g_m(\cap_n (g_n(B))) \text{ by (1.1)} \\
&= \cap_{B \in \mathcal{D}} g_m(B) \text{ by (a).}
\end{aligned}$$

This completes the proof. $\square$

Recall that the class $\mathcal{A}(u)$ consists of all finite unions of sets from $\mathcal{A}$. We shall define the semi-algebra $\mathcal{C}$ to be the class of all subsets of T of the form
$$C = A \setminus B,\ A \in \mathcal{A},\ B \in \mathcal{A}(u).$$

FRAMEWORK AND ASSUMPTIONS   13

Denote by $\mathcal{C}_n$ the subset of $\mathcal{C}$ of sets of the form
$$C = A \setminus B, \ A \in \mathcal{A}_n, \ B \in \mathcal{A}_n(u).$$

$\mathcal{C}$ is closed under intersections and any set in $\mathcal{C}(u)$ (the set of finite unions of sets in $\mathcal{C}$) may be expressed as a finite disjoint union of sets in $\mathcal{C}$. Note that if $B = \cup_{i=1}^k A_i \in \mathcal{A}(u)$, without loss of generality we can require that for each i, $A_i \not\subseteq \cup_{j \neq i} A_j$. Such a representation of $B \in \mathcal{A}(u)$ will be called *extremal*. If $C = A \setminus B$, $A \in \mathcal{A}$, $B \in \mathcal{A}(u)$, then the representation of C is called extremal if that of B is. Unless otherwise stated, it will always be assumed that all representations of sets in $\mathcal{A}(u)$ and C are extremal.

The following assumption about the structure of $\mathcal{C}$ will be required.

Assumption 1.1.4 *For any $B \in \mathcal{A}(u)$, if $C = A \setminus \cup_{i=1}^k A_i \in \mathcal{C}$ and if $A \subseteq B$, then there exist sets $D_1, ..., D_m \in \mathcal{A}$, $D_i \subseteq B$, $i = 1, ..., m$ such that $C = A \setminus \cup_{i=1}^m D_i$ is an extremal representation, and if $A' \in \mathcal{A}$, $A' \subseteq B$, $A' \cap C = \emptyset$, then $A' \subseteq \cup_{i=1}^m D_i$. This is called a* maximal *representation of C in B.*

The next assumption will be used in some situations, but is not imposed unless specifically stated. It is labelled and will be referred to henceforth as 'SHAPE' as it imposes a restriction on the geometric shapes of the sets in $\mathcal{A}$.

Assumption 1.1.5 (SHAPE) *If $A, A_1, ..., A_n \in \mathcal{A}$ and $A \subseteq \cup_{i=1}^n A_i$, then there is an index i, $1 \leq i \leq n$ such that $A \subseteq A_i$.*

Comments:
1. Without loss of generality (cf. Definition 1.1.1 4(a)) it may be assumed that the finite sub-semilattices $\mathcal{A}_n$ satisfy SHAPE whenever this property is required.
2. Note that when $\mathcal{A}$ satisfies SHAPE, all nonempty sets $B \in \mathcal{A}(u)$ and $C \in \mathcal{C}$ have unique extremal representations $B = \cup_{i=1}^k A_i$, $A_1, ..., A_k \in \mathcal{A}$ and $C = A \setminus B$, $A \in \mathcal{A}$, $B \in \mathcal{A}(u)$, $B \subseteq A$.

Lemma 1.1.6 *Assume that $\mathcal{A}$ satisfies SHAPE. If $C = A \setminus \cup_{i=1}^k A_i \neq \emptyset$ is an extremal representation of $C \in \mathcal{C}$, then for each $i = 1, ..., k$ and $n = 1, 2, ...$, $g_n(A_i) \cap C \neq \emptyset$.*

Proof. Without loss of generality, it may be assumed that $A_i \subset A$ $\forall i, 1 \leq i \leq k$. Then if $g_n(A_i) \cap C = \emptyset$, by Comment #5 following Definition 1.1.1, $A_i \subset g_n(A_i) \cap A \subseteq \cup_{j=1}^k A_j$. By SHAPE it follows that there exists $j \neq i$ such that $g_n(A_i) \cap A \subseteq A_j$. Thus $A_i \subset$

A_j, contradicting the assumption that the representation of C is extremal. □

Comment: In all of our examples, including those in which SHAPE is not satisfied, it is easy to see that if $C = A \backslash B$, $A \in \mathcal{A}$, $B \in \mathcal{A}(u)$, then $g_n(B) \cap C \neq \emptyset$, $\forall n$. Henceforth, we shall always assume that this is the case.

For certain results, we require additional assumptions, which will be imposed as needed. Let $\mathcal{A}'$ be any finite sub-semilattice of $\mathcal{A}$. The class $\mathcal{C}^\ell(\mathcal{A}')$ of *left-neighbourhoods* of $\mathcal{A}'$ is the class of all sets in $\mathcal{C}$ of the form

$$C = A \setminus \cup_{A' \in \mathcal{A}', A \not\subseteq A'} A' = A \setminus \cup_{A' \in \mathcal{A}', A' \subset A} A', A \in \mathcal{A}'.$$

The set C in the expression above is known as the left-neighbourhood of A in $\mathcal{A}'$.

Assumption 1.1.7 *Let $\mathcal{C}^\ell(\mathcal{A}_n)$ denote the class of left neighbourhoods of $\mathcal{A}_n$. The sequence $\{\mathcal{C}^\ell(\mathcal{A}_n)\}_n$ is a dissecting system for T (in other words, if $s, t \in T, s \neq t$, then $\exists n, C, C' \in \mathcal{C}^\ell(\mathcal{A}_n)$ such that $s \in C$, $t \in C'$ and $C \cap C' = \emptyset$).*

In many applications, we will have to number the sets in a finite sub-semilattice $\mathcal{A}'$ of $\mathcal{A}$ in a particular way.

Definition 1.1.8 *Given a finite sub-semilattice $\mathcal{A}'$ of $\mathcal{A}$, and sets $A, B \in \mathcal{A}'$, we shall say that B is minimal (in $\mathcal{A}'$) with respect to A if $A \subset B$ and $\not\exists D \in \mathcal{A}'$ such that $A \subset D \subset B$. (Recall that '$\subset$' denotes strict inclusion.)*

Let $\mathcal{A}'$ be any finite sub-semilattice of $\mathcal{A}$. The sets in $\mathcal{A}'$ can always be numbered in the following way: $A_0 = \emptyset'$, and given $A_0, ..., A_{i-1}$, choose A_i to be any set in $\mathcal{A}'$ such that $A \subset A_i$ implies that $A = A_j$, some $j = 1, ..., i-1$. Any such numbering $\mathcal{A}' = \{A_0, ..., A_k\}$ will be called 'consistent with the strong past' (i.e., if C_i is the left-neighbourhood of A_i in $\mathcal{A}'$, then $C_i = \cup_{j=0}^{i} A_j \setminus \cup_{j=0}^{i-1} A_j$ and $A_j \cap C_i = \emptyset$, $\forall j = 0, ..., i-1$, $i = 1, ..., k$).

Lemma 1.1.9 *Let $\mathcal{A}'$ be any finite sub-semilattice of $\mathcal{A}$. Then for any numbering $\mathcal{A}' = \{A_0, ..., A_k\}$ consistent with the strong past and for each $i, 1 \leq i \leq k$, there exists $j_i < i$ such that A_i is minimal (in $\mathcal{A}'$) with respect to A_{j_i}.*

Proof. Given any finite sublattice $\mathcal{A}' = \{A_0, ..., A_k\}$, equipped with a numbering consistent with the strong past, suppose there exists $A_i, i \geq 1$ such that no index $j_i < i$ exists such that A_i is minimal

EXAMPLES

(in $\mathcal{A}'$) with respect to A_{j_i}. Then we may choose $j_1 < i$ such that $A_{j_1} \subset A_i$, and since A_i is not minimal with respect to A_{j_1}, there exists $B \in \mathcal{A}'$ such that $A_{j_1} \subset B \subset A_i$. Since the numbering of $\mathcal{A}'$ is consistent with the strong past, it follows that $B = A_{j_2}$ for some j_2, $j_1 < j_2 < i$. But by assumption, A_i is not minimal with respect to A_{j_2}, so there exists j_3, $j_1 < j_2 < j_3 < i$ such that $A_{j_1} \subset A_{j_2} \subset A_{j_3} \subset A_i$. Continuing in this way, there exists a strictly increasing sequence $j_1 < j_2 < ... < i$ such that $A_{j_1} \subset A_{j_2} \subset ... \subset A_i$, which contradicts the assumption that $\mathcal{A}'$ is finite. This completes the proof. $\square$

1.2 Examples

In this section we introduce various examples of topological spaces T and indexing collections $\mathcal{A}$.

Example 1.2.1 The classical example is $T = \mathbf{R}_+^d$ and $\mathcal{A} = \{[0, \mathbf{x}] : \mathbf{x} \in \mathbf{R}_+^d\} \cup \{\emptyset\}$ (see Figure 1.1). We may define $B_n = [0, \mathbf{n}]$ and $\mathcal{A}_n = \{[0, \mathbf{x}] : x_i = k_i/2^n, 0 \leq k_i \leq n2^n, i = 1, ..., d\} \cup \{\emptyset\}$. Then $g_n(\emptyset) = \emptyset$, $g_n([0, \mathbf{x}]) = [0, (k_1/2^n, ..., k_d/2^n)]$, where k_i is such that $(k_i - 1)/2^n \leq x_i < k_i/2^n$, $i = 1, ..., d$ for every $\mathbf{x} = (x_1, ..., x_d)$ such that $x_i < n, i = 1, ..., d$, and $g_n([0, \mathbf{x}]) = \{T\}$ otherwise. We have $\emptyset' = \{\mathbf{0}\}$ and that $\tilde{\mathcal{A}}(u)$ is the class of all 'lower sets' as defined in Example 1.2.4 below. Note that $\mathcal{A}$ is a distributive lattice, that SHAPE is satisfied and that $g_n : \mathcal{A} \to \mathcal{A}_n \cup \{T\}$. The left neighbourhoods $\mathcal{C}^\ell(\mathcal{A}_n)$ are the $1/2^n$-sided squares in B_n, and clearly the sequence $(\mathcal{C}^\ell(\mathcal{A}_n))_n$ is a dissecting system for T. As well, Assumption 1.1.4 is satisfied.

Example 1.2.2 The preceding example is very close to the one-dimensional astrophysical model suggested by P. Greenwood. We have a one-dimensional universe which begins at 0 at time 0 with a 'big bang'. Matter and energy then begin to spread in both directions from the origin with a maximum speed of one unit of distance per unit of time. As we observe the evolution of our one-dimensional universe, we see that the set of locations and times may be defined by $T = \{(x, t) \in \mathbf{R} \times \mathbf{R}_+ : |x| \leq t\}$ (see Figure 1.2). We now define the set $A(x, t)$ to be the intersection with T of the triangle with vertices $(x, t), (x - t, 0), (x + t, 0)$ and we let $\mathcal{A} = \{A(x, t) : (x, t) \in T\} \cup \{\emptyset\}$. The sets in $\mathcal{A}$ and $\mathcal{C}$ have simple interpretations: an observer at $x \in \mathbf{R}$ can see an event occurring at location $y \in \mathbf{R}$ at time s by time t if and only if $(y, s) \in A(x, t)$. If

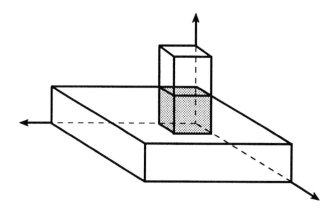

Figure 1.1

$C = A \setminus \cup_{i=1}^{k} A_i \in \mathcal{C}$, then events occurring at locations and times in C are those which are visible to an observer at x by time t, but are not visible to observers at locations and times in $\cup_{i=1}^{k} A_i$. It is easily seen that, just as in the preceding example, all the assumptions of Section 1.1 are satisfied.

Example 1.2.3 Example 1.2.1 can be extended to $T = \mathbf{R}^d$ as follows: $\mathcal{A}$ consists of all sets of the form $A = \emptyset$ or $A = Q_i$ where $Q_i = [\mathbf{0}, \mathbf{x_i}]$, and $\mathbf{x_i}$ is a point in the ith closed quadrant of $\mathbf{R}^d$. We now have a sort of 2^d-sided process. $B_n = [-\mathbf{n}, \mathbf{n}]$ and the class $\mathcal{A}_n$ now consists of $\emptyset$ and all sets in $\mathcal{A}$ of the form $[\mathbf{0}, \mathbf{x}] : x_i = k_i/2^n, 0 \le \mid k_i \mid \le n2^n, i = 1, ..., d$. The functions $g_n : \mathcal{A} \to \mathcal{A}_n(u) \cup T$ since we must have $A \subseteq g_n(A)^\circ$. For example, if $d = 2$ and $\mathbf{x}$ is such that $0 \le x_i < n, i = 1, 2$ then $g_n([\mathbf{0}, \mathbf{x}]) = [\mathbf{0}, (k_1/2^n, k_2/2^n)] \cup [\mathbf{0}, (-1/2^n, k_2/2^n)] \cup [\mathbf{0}, (k_1/2^n, -1/2^n)] \cup [\mathbf{0}, (-1/2^n, -1/2^n)]$, where k_i is such that $k_{i-1}/2^n \le x_i < k_i/2^n$, $i = 1, 2$. As before, the sets in $\mathcal{C}^\ell(\mathcal{A}_n)$ are the $1/2^n$-sided squares in B_n, and the sequence $(\mathcal{C}^\ell(\mathcal{A}_n))_n$ is a dissecting system for T. SHAPE is still satisfied in this example.

EXAMPLES

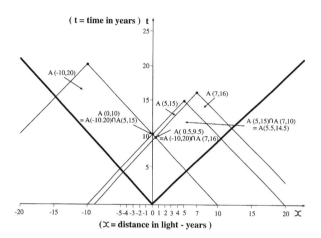

Figure 1.2

Example 1.2.4 Examples 1.2.1 and 1.2.3 may be generalized as follows. Let $T = \mathbf{R}_+^d$ or $\mathbf{R}^d$ and take $\mathcal{A}$ to be the class of compact *lower sets*: i.e., the class of compact subsets A of T satisfying $\mathbf{t} \in A$ implies $[\mathbf{0}, \mathbf{t}] \subseteq A$. SHAPE is not satisfied. Note that in this case $\mathcal{A} = \mathcal{A}(u) = \bar{\mathcal{A}}(u)$, and $\mathcal{A}$ is a lattice. The sets B_n may be defined as before, and the semilattices $\mathcal{A}_n$ will now consist of finite unions of sets of the form $[\mathbf{0}, \mathbf{x}] : x_i = k_i/2^n, 0 \leq |k_i| \leq n2^n, i = 1, ..., d$. If $A \subseteq B_n^\circ$, then $g_n(A) = \cap_{A' \in \mathcal{A}_n, A \subseteq (A')^\circ} A'$; otherwise $g_n(A) = T$.

Example 1.2.5 It is easily seen that Example 1.2.4 may be modified so that $\mathcal{A}$ includes only the convex lower sets (see Figure 1.3).

Example 1.2.6 This is an example in which T is a compact subset of the function space $D[0, 1]$ of functions $f : [0, 1] \to \mathbf{R}$ which are right continuous with left limits at every point (and continuous from the left at 1). Consider a finite decomposition $I_0, I_1, ..., I_N$ of $[0, 1]$ where $0 = d_0 < d_1 < ... < d_N < 1$ and $I_k = [d_k, d_{k+1})$ if $k < N$ and $I_N = [d_N, 1]$. Let

$$T = \{f \in D[0,1] : 0 \leq f(x) \leq 1 \; \forall x \in [0,1], \text{ and } \forall \epsilon > 0,$$

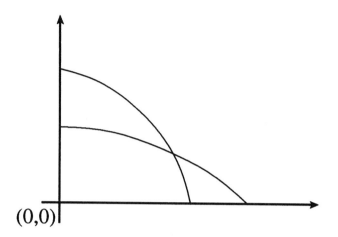

(0,0)

Figure 1.3

$| f(x) - f(y) | < \epsilon$ $\forall x, y$ such that $| x - y | < \epsilon$ and $x, y \in I_i$, some $i = 0, ..., N$ }.

T is a compact subset of $D[0, 1]$ under the sup norm topology. The functions in T may have discontinuities at $d_0, ..., d_N$ and are uniformly equicontinuous on $I_0, ..., I_N$. There are various ways in which an indexing collection may be defined on T, but we shall define $\mathcal{A}$ as follows: let H denote the class of functions of the form $h = \sum_{i=0}^{N} h_i \chi_{I_i}$, where χ denotes the indicator function and (h_i) is a sequence of constants such that $0 \leq h_i \leq 1$ $\forall i$. Then for $h \in H$, let $A_h = \{f \in T : f(x) \leq h(x) \; \forall x \in [0, 1]\}$. Finally, let $\mathcal{A} = \{A_h : h \in H\} \cup \emptyset$.

$\mathcal{A}$ is a distributive lattice under the following operations:

$$A_g \wedge A_h = A_g \cap A_h = A_{g \wedge h}$$

where $g \wedge h = \sum (g_i \wedge h_i) \chi_{I_i}$, and

$$A_g \vee A_h = A_{g \vee h}$$

where $g \vee h = \sum (g_i \vee h_i) \chi_{I_i}$. Clearly $A_g \cup A_h \subseteq A_g \vee A_h$, but the two sets are not equal in general and $\mathcal{A} \neq \mathcal{A}(u)$. We observe that A_h° consists of all $f \in A_h$ for which there exists $\epsilon > 0$ such that $f(x) \leq h(x) - \epsilon$ whenever $h(x) < 1$. If $h_i = 0$ for any i then $A_h^\circ = \emptyset$. We have that $\emptyset' = \{0\}$ and that $\mathcal{A}$ satisfies SHAPE.

The sublattices $\mathcal{A}_n$ are defined as follows:

$$\mathcal{A}_n = \{A_h \in \mathcal{A} : h = \sum_{i=0}^{N} h_i \chi_{I_i}, \ h_i \in D_n, \ i = 0, ..., N\} \cup \{\emptyset\},$$

where D_n denotes the dyadics of order n. For $h = \sum_{i=0}^{N} h_i \chi_{I_i} \in H$, define $g_n : \mathcal{A} \to \mathcal{A}_n$ as follows:

$$g_n(\emptyset) = \emptyset,$$

$$g_n(A_h) = \sum_{i=0}^{N} (h_i)^{n+} \chi_{I_i},$$

where $(h_i)^{n+} = (k+1)/2^n$ if $k/2^n \leq h_i < (k+1)/2^n$ and $(h_i)^{n+} = 1$ if $h_i = 1$.

This is the first example which does not satisfy Assumption 1.1.7: $(\mathcal{C}^\ell(\mathcal{A}_n))_n$ is not a dissecting system for T. To see this, observe that the function which is identically equal to 1 cannot be separated by left-neighbourhoods from any other function f in T for which there exists a point $x_i \in I_i$ such that $f(x_i) = 1$, for each $i \geq 0$. If we were to restrict the functions in T to functions which remained constant on each I_i, Assumption 1.1.7 would be satisfied.

1.3 The Hausdorff Metric

This section will be used primarily in Part 2 of the book, and thus may be skipped on first reading and referred to as necessary.

Different topologies may be defined on families of subsets of T. In this section, we will assume that T is a complete separable metric space with metric d, which is the case with all the examples of the preceding section. We will consider the *Hausdorff topology* defined on $\mathcal{K}\setminus\emptyset$.

For any $A \subseteq T$, $A \neq \emptyset$ and $t \in T$, denote

$$d(A, t) = \inf\{d(s, t) : s \in A\},$$

$$A^\epsilon = \{t \in T : d(A, t) \leq \epsilon\}, \text{ for any } \epsilon > 0.$$

We define $\emptyset^\epsilon = \emptyset$. Now define the Hausdorff distance between two nonempty sets $A, B \subseteq T$ by

$$d_H(A, B) = \inf\{\epsilon > 0 : A \subseteq B^\epsilon \text{ and } B \subseteq A^\epsilon\}.$$

On the class $\mathcal{K}\setminus\emptyset$, d_H is a metric called the Hausdorff metric.

The proof of the following result on monotone sequences is a straightforward exercise:

Lemma 1.3.1 *If T is compact, then for any increasing (decreasing) sequence (B_n) of closed sets, $B = \overline{\cup_n B_n}$, $(B = \cap_n B_n)$ if and only if B is the limit of (B_n) in the Hausdorff topology.*

When T is a metric space, we will make the following assumption about our indexing collection $\mathcal{A}$, which is easily seen to be satisfied by all of our examples:

Assumption 1.3.2 *The sub-semilattices $\mathcal{A}_n$ and the functions g_n may be chosen so that there exists a sequence (ϵ_n), $\epsilon_n \downarrow 0$ such that for each $n \in \mathbf{N}$,*

$$d_H(A, g_n(A)) \leq \epsilon_n, \ \forall A \in \mathcal{A} \text{ such that } g_n(A) \subseteq B_n.$$

Comment: It is a straightforward observation that $\mathcal{A}$ may be replaced by $\mathcal{A}(u)$ in Assumption 1.3.2.

It is well known that if (T, d) is compact, then so is $(\mathcal{K} \setminus \emptyset, d_H)$. The following lemma from Slonowsky (1998) states that any indexing collection $\mathcal{A}$ satisfying Assumption 1.3.2 is d_H closed; thus, if (T, d) is compact, then so is $(\mathcal{A} \setminus \emptyset, d_H)$.

Lemma 1.3.3 *If $\mathcal{A}$ is an indexing collection on T satisfying Assumption 1.3.2 and if $g_n : \mathcal{A} \to \mathcal{A}_n \cup T$, then $\mathcal{A}$ is d_H-closed in $\mathcal{K}$.*

Proof. Suppose that $A_n \xrightarrow{d_H} K$, $A_n \in \mathcal{A} \ \forall n$ and $K \in \mathcal{K}$. We first observe that since K is compact, Definition 1.1.1 #1 ensures that there exists N such that $A_n \subseteq B_N$ for every n. Thus, in what follows we shall assume that $n \geq N$. Now, #4 of Definition 1.1.1 ensures that there is a sequence (δ_n), $\delta_n \downarrow 0$ such that $A^{\delta_n} \subseteq g_n(A)$, $\forall A \in \mathcal{A}_n$.

Now, it must be shown that $K \in \mathcal{A}$. Using the sequence (δ_n) defined above, we construct a subsequence (A_{k_n}) as follows: given $n \in \mathbf{N}$, choose $k_n \in \mathbf{N}$ such that $d_H(A_{k_n}, K) < \delta_n$. Without loss of generality, we may suppose that $(k_n)_n$ is strictly increasing in n. Therefore,

$$K \subseteq A_{k_n}^{\delta_n} \subseteq [g_n(A_{k_n})]^{\delta_n} \subseteq g_n^2(A_{k_n}), \ \forall n,$$

where g^2 denotes $g \circ g$. Moreover, given any $n \geq N$, Assumption 1.3.2 implies

THE HAUSDORFF METRIC

$$d_H(g_n^2(A_{k_n}), K) \leq d_H(g_n^2(A_{k_n}), g_n(A_{k_n})) + d_H(g_n(A_{k_n}), A_{k_n})$$
$$+ d_H(A_{k_n}, K)$$
$$\leq \epsilon_n + \epsilon_n + \delta_n$$

where $2\epsilon_n + \delta_n \to 0$ as $n \to \infty$. Thus, $d_H(g_n^2(A_{k_n}), K) \to 0$, and since $K \subseteq g_n^2(A_{k_n}) \; \forall n$, it is straightforward to see that $K = \cap_n g_n(A_{k_n})$. Since $\mathcal{A}$ is closed under countable intersections, $K \in \mathcal{A}$, as required. $\square$

Although in some of our examples $g_n : \mathcal{A} \to \mathcal{A}_n(u) \cup T$, the preceding lemma still holds as it is easily seen that for these examples $K = \cap_n g_n(A_{k_n}) \in \mathcal{A}$ for any Cauchy sequence $(A_{k_n})_n$ in $\mathcal{A}$.

When we impose a natural assumption on $\mathcal{A}_n$ ensuring that all sets in $\tilde{\mathcal{A}}(u)$ may be approximated uniformly from above and below by sets in $\mathcal{A}_n(u)$, we gain more insight into the structure of sets in $\tilde{\mathcal{A}}(u)$, as is shown in the following lemma:

Lemma 1.3.4 *If for each $B \in \tilde{\mathcal{A}}(u)$ and each $\epsilon > 0$ there exists $m = m_{B,\epsilon}$ such that $d_H(g_n(B), B_n^-) < \epsilon$, $\forall n \geq m$, where*

$$B_n^- = \bigcup_{\substack{B' \in \mathcal{A}_n(u) \\ B' \subseteq B}} B',$$

then each set $B \in \tilde{\mathcal{A}}(u)$ may be expressed as the closure of a countable union of sets in $\mathcal{A}$: i.e.

$$\overline{\bigcup_n B_n^-} = \cap_n g_n(B) = B.$$

Proof. We have $\overline{\cup_n B_n^-} \subseteq \cap_n g_n(B) = B$. Conversely, suppose that $y \in \cap_n g_n(B) = B$. For $k \in \mathbb{N}$, $\exists m_k$ such that $d(y, B_{m_k}^-) < 1/k$, and so $\exists y_k \in B_{m_k}^-$ satisfying $d(y, y_k) < 1/k$. Thus, we may choose a sequence $(y_k) \in \cup_n B_n^-$ such that $y_k \to y$, and so $y \in \overline{\cup_n B_n^-}$. $\square$

All of our examples satisfy the condition of Lemma 1.3.4.

We close this section with a proposition from Slonowsky (1998) which illustrates some restrictions on the cardinality of the indexing collection.

Proposition 1.3.5 *Let (T, d) be a compact metric space. If $\mathcal{A}$ is an indexing collection on T, then*
(a) the cardinality of $\mathcal{A}$ cannot exceed that of $\mathcal{P}(\mathbb{R})$ (the power set of $\mathbb{R}$).
(b) If in addition (T, d) is connected, then $\mathcal{A}$ is either infinite or $\mathcal{A} = \{\emptyset, T\}$.

Proof.
(a): It is a well known result of point-set topology that the cardinality of any set endowed with a separable Hausdorff topology cannot exceed that of $\mathcal{P}(\mathbf{R})$. The result follows, since $(\mathcal{A}, d_H)$ is Hausdorff and $\mathcal{A}' = \cup_n \mathcal{A}_n$ is a countable d_H-dense subset of $\mathcal{A}$.
(b): Assuming that (T, d) is connected, suppose that $\mathcal{A}$ is finite and that there exists $A \in \mathcal{A}$ such that $A \neq \emptyset$ and $A \neq T$. Since $A = \cap_n g_n(A)$, there exists N such that $g_n(A) \neq T\ \forall n \geq N$. Since T is connected,
$$(g_n(A))^\circ \subset g_n(A),\ \forall n \geq N.$$
However, by the assumption that $\mathcal{A}$ is finite, $g_n(A) = A$ for all n sufficiently large. Hence, there exists $n \geq N$ such that
$$g_n(A) = A \subseteq (g_n(A))^\circ \subset g_n(A),$$
which leads to a contradiction. Thus, $\mathcal{A}$ must be infinite. □

1.4 The Probability Structure

In this section we describe the probability structure required for the development of martingale theory.

Let $(\Omega, \mathcal{F}, P)$ be a complete probability space equipped with an $\mathcal{A}$-indexed filtration $\{\mathcal{F}_A : A \in \mathcal{A}\}$ which satisfies the following conditions:

- $\forall A \in \mathcal{A}$, we have $\mathcal{F}_A \subseteq \mathcal{F}$ and $\mathcal{F}_A$ contains the P-null sets.
- $\forall A, B \in \mathcal{A}$, $\mathcal{F}_A \subseteq \mathcal{F}_B$, if $A \subseteq B$.
- Monotone outer-continuity: $\mathcal{F}_{\cap A_i} = \cap \mathcal{F}_{A_i}$ for any decreasing sequence (A_i) in $\mathcal{A}$.
- For consistency in what follows, if $T \notin \mathcal{A}$, define $\mathcal{F}_T = \mathcal{F}$.

We may associate various σ-algebras with sets in $\mathcal{A}(u)$, $\tilde{\mathcal{A}}(u)$ and $\mathcal{C}(u)$. If $B \in \tilde{\mathcal{A}}(u)$, then $\mathcal{F}_B^0 = \vee_{A \in \mathcal{A}, A \subseteq B} \mathcal{F}_A$. The σ-algebras $\{\mathcal{F}_B^0 : B \in \tilde{\mathcal{A}}(u)\}$ are complete and increasing, but not necessarily monotone outer-continuous. Thus, we define for $B \in \tilde{\mathcal{A}}(u)$: $\mathcal{F}_B^r = \cap_n \mathcal{F}_{g_n(B)}^0$.

Lemma 1.4.1 *(Ivanoff and Merzbach (1994))* Let (B_n) be a decreasing sequence in $\tilde{\mathcal{A}}(u)$. If $B = \cap_n B_n$, then $\mathcal{F}_B^r = \cap_n \mathcal{F}_{B_n}^r$.

Proof. Clearly, $\mathcal{F}_B^r \subseteq \cap_n \mathcal{F}_{B_n}^r$. Conversely, if $F \in \cap_n \mathcal{F}_{B_n}^r$, then $F \in \cap_n \cap_m \mathcal{F}_{g_m(B_n)}^0$. But for m fixed, by Lemma 1.1.3 (b), $g_m(B_n) =$

THE PROBABILITY STRUCTURE

$g_m(B)$ for all n sufficiently large, since $\mathcal{A}_m$ is finite. Therefore, $F \in \cap_m \mathcal{F}^0_{g_m(B)} = \mathcal{F}^r_B$. □

At times we will require monotone outer-continuity, at other times not, although unless otherwise stated *it will always be assumed that*

- $\mathcal{F}_A = \cap_n \mathcal{F}_{g_n(A)}$ for all $A \in \mathcal{A}$.

For $B \in \tilde{\mathcal{A}}(u)$, in the sequel $\mathcal{F}_B$ will denote either $\mathcal{F}^0_B$ or $\mathcal{F}^r_B$ as required.

If $C \in \mathcal{C}(u) \setminus \mathcal{A}$, define $\mathcal{G}_C = \cap_{A \in \mathcal{A}, A \cap C \neq \emptyset} \mathcal{F}_A$. Next, for $C \in \mathcal{C}(u) \setminus \mathcal{A}$, let $\mathcal{G}^*_C = \vee_{B \in \mathcal{A}(u), B \cap C = \emptyset} \mathcal{F}_B$. Trivially, if $B \in \mathcal{A}(u)$ and $B \cap C = \emptyset$, then $\mathcal{F}_B \subseteq \mathcal{G}^*_C$. Thus, it is an easy consequence of monotone outer-continuity of the filtration $\{\mathcal{F}_A : A \in \mathcal{A}\}$ and the comment following Lemma 1.1.6 that $\mathcal{G}_C \subseteq \mathcal{G}^*_C$ for $C \in \mathcal{C} \setminus \mathcal{A}$. For consistency, therefore, we define $\mathcal{G}^*_A = \mathcal{G}_A = \mathcal{F}_\emptyset$ for $A \in \mathcal{A}$.

Note that both $\{\mathcal{G}_C\}$ and $\{\mathcal{G}^*_C\}$ are decreasing families of σ-algebras. $\{\mathcal{G}_C\}$ and $\{\mathcal{G}^*_C\}$ will be referred to as the *weak* and *strong* histories of C, respectively. To illustrate with Example 1.2.1, let $T = \mathbf{R}^2_+$ and $\mathcal{A} = \{[\mathbf{0}, \mathbf{x}] : \mathbf{x} \in T\}$, and denote $\mathcal{F}_{[\mathbf{0}, \mathbf{x}]} = \mathcal{F}_\mathbf{x}$. Now, if

$$C = (\mathbf{y}, \mathbf{x}] = [(0,0), (x_1, x_2)] \setminus [(0,0), (y_1, x_2)] \cup [(0,0), (x_1, y_2)],$$

it follows by outer-continuity that $\mathcal{G}_C = \mathcal{F}_\mathbf{y}$ and that

$$\mathcal{G}^*_C = \vee_{z_1 \leq y_1 \text{ or } z_2 \leq y_2} \mathcal{F}_{(z_1, z_2)}.$$

Occasionally we shall impose the following condition of commutation on the filtration:

Definition 1.4.2 (CI) *The filtration $\{\mathcal{F}_A; A \in \mathcal{A}\}$ is said to satisfy the conditional independence property (CI) if for all $A, B \in \mathcal{A}$, the conditional expectation operator $E[\cdot \mid \cdot]$ satisfies*

$$E[E[\cdot \mid \mathcal{F}_A] \mid \mathcal{F}_B] = E[\cdot \mid \mathcal{F}_{A \cap B}],$$

or equivalently, $\mathcal{F}_A$ and $\mathcal{F}_B$ are conditionally independent given $\mathcal{F}_{A \cap B}$ (denoted by $\langle \mathcal{F}_A \perp \mathcal{F}_B \mid \mathcal{F}_{A \cap B} \rangle$).

The equivalence of the two conditions above is an easy consequence of the well-known fact that if $\mathcal{F}$, $\mathcal{G}$ and $\mathcal{H}$ are σ-algebras such that $\mathcal{H} \subseteq \mathcal{G}$, then $\mathcal{F}$ and $\mathcal{G}$ are conditionally independent given $\mathcal{H}$ if and only if for each $\mathcal{F}$-measurable random variable X, $E[X \mid \mathcal{G}] = E[X \mid \mathcal{H}]$. (CI) corresponds to the (F4) condition introduced by Cairoli and Walsh (1975) which became a routine assumption in the literature on martingales with a multi-dimensional time parameter. It is easily seen that (CI) implies that $\mathcal{F}_A \cap \mathcal{F}_B = \mathcal{F}_{A \cap B}$.

We have the following lemma from Slonowsky (1998):

Lemma 1.4.3 *Assume that (CI) holds and that there exists a binary operation $\vee$ on $\mathcal{A}$ such that $\mathcal{A}$ is a distributive lattice under $\vee$ and $\wedge = \cap$. If $A, A_1, ..., A_n \in \mathcal{A}$ are such that $A_i \cap A_j \subseteq A$ $\forall i \neq j$, then for any $X \in L^1(\mathcal{F}_A)$,*

$$E[X \mid \vee_{i=1}^n \mathcal{F}_{A_i}] = E[X \mid \vee_{i=1}^n \mathcal{F}_{A \cap A_i}].$$

Proof. We begin with the observation that for any sub-σ-algebras $\mathcal{F}_1, \mathcal{F}_2, \mathcal{F}_3, \mathcal{F}_4$ of $\mathcal{F}$, if $\langle \mathcal{F}_1 \perp \mathcal{F}_2 \mid \mathcal{F}_3 \rangle$ and $\mathcal{F}_4 \subseteq \mathcal{F}_2$, then

$$\langle (\mathcal{F}_1 \vee \mathcal{F}_4) \perp \mathcal{F}_2 \mid (\mathcal{F}_3 \vee \mathcal{F}_4) \rangle. \tag{1.2}$$

This is a straightforward exercise using properties of conditional independence, and is proven in Appendix B of Slonowsky (1998).

Now, suppose $1 \leq j \leq n$. Since $\mathcal{A}$ is a distributive lattice and $A_j \cap A_i \subseteq A_j \cap A$ $\forall i \neq j$, $A_j \cap [A \vee (\vee_{i=j+1}^n A_i)] = A_j \cap A$, which implies, by (CI), that

$$\langle \mathcal{F}_{A_j} \perp \mathcal{F}_{A \vee (\vee_{i=j+1}^n A_i)} \mid \mathcal{F}_{A \cap A_j} \rangle.$$

By (1.2) with $\mathcal{F}_4 = (\vee_{i=1}^{j-1} \mathcal{F}_{A \cap A_i}) \vee (\vee_{i=j+1}^n \mathcal{F}_{A_i})$, it follows that

$$\langle ((\vee_{i=1}^{j-1} \mathcal{F}_{A \cap A_i}) \vee (\vee_{i=j}^n \mathcal{F}_{A_i}) \perp \mathcal{F}_{A \vee (\vee_{i=j+1}^n A_i)}$$
$$\mid (\vee_{i=1}^j \mathcal{F}_{A \cap A_i}) \vee (\vee_{i=j+1}^n \mathcal{F}_{A_i}) \rangle$$

which in turn implies

$$\langle ((\vee_{i=1}^{j-1} \mathcal{F}_{A \cap A_i}) \vee (\vee_{i=j}^n \mathcal{F}_{A_i}) \perp \mathcal{F}_A$$
$$\mid (\vee_{i=1}^j \mathcal{F}_{A \cap A_i}) \vee (\vee_{i=j+1}^n \mathcal{F}_{A_i}) \rangle. \tag{1.3}$$

(Clearly, $\vee_{i=1}^{j-1}$ and $\vee_{i=j+1}^n$ are null for $j = 1$ and $j = n$, respectively.) Finally, take $X \in L^1(\mathcal{F}_A)$. Equation (1.3) implies that for $1 \leq j \leq n$,

$$E[X \mid (\vee_{i=1}^{j-1} \mathcal{F}_{A \cap A_i}) \vee (\vee_{i=j}^n \mathcal{F}_{A_i})] =$$
$$E[X \mid (\vee_{i=1}^j \mathcal{F}_{A \cap A_i}) \vee (\vee_{i=j+1}^n \mathcal{F}_{A_i})]. \tag{1.4}$$

Applying iteratively the n identities defined by equation (1.4), we obtain

$$E[X \mid \vee_{i=1}^n \mathcal{F}_{A_i}] = E[X \mid \vee_{i=1}^n \mathcal{F}_{A \cap A_i}],$$

completing the proof. □

THE PROBABILITY STRUCTURE

We end this section with a short discussion of $\mathcal{A}$-indexed processes.

Definition 1.4.4 *A ($\mathcal{A}$-indexed) stochastic process $X = \{X_A : A \in \mathcal{A}\}$ is a collection of random variables indexed by $\mathcal{A}$, and is said to be adapted if X_A is $\mathcal{F}_A$-measurable, for every $A \in \mathcal{A}$. X is said to be integrable (square integrable) if $E[\|X_A\|] < \infty$ ($E[(X_A)^2] < \infty$) for every $A \in \mathcal{A}$. More generally, a process X is said to be an L^p process if $E[\|X_A\|^p] < \infty$ for every $A \in \mathcal{A}$. It will always be assumed that $X_\emptyset = 0$.*

Comment: Various authors have considered processes indexed by lattices. To see that our framework includes such processes, assume that T is a complete lattice, and for $t \in T$, let $A_t = \{s \in T : s \leq t\}$. Since $A_t \cap A_{t'} = A_{t \wedge t'}$, we may define $\mathcal{A} = \{A_t : t \in T\}$ and note that $\mathcal{A}$ satisfies SHAPE. We may identify a T-indexed process X with the set-indexed process $X_{A_t} := X_t$.

Any $\mathcal{A}$-indexed function which has a (finitely) additive extension to $\mathcal{C}$ will be called *additive* (and is easily seen to be additive on $\mathcal{C}(u)$ as well). This is always the case if SHAPE holds, as in the lattice example of the preceding comment. For stochastic processes, we do not necessarily require that each sample path be additive, but additivity will be imposed in an almost sure sense:

Definition 1.4.5 *A ($\mathcal{A}$-indexed) stochastic process X is additive if it has an (almost sure) additive extension to $\mathcal{C}$: i.e., $X_\emptyset = 0$ and if $C, C_1, C_2 \in \mathcal{C}$ with $C = C_1 \cup C_2$ and $C_1 \cap C_2 = \emptyset$, then almost surely*

$$X_{C_1} + X_{C_2} = X_C.$$

In particular, if $C \in \mathcal{C}$ and $C = A \setminus \cup_{i=1}^n A_i$, $A, A_1, ..., A_n \in \mathcal{A}$, then almost surely

$$X_C = X_A - \sum_{i=1}^n X_{A \cap A_i} + \sum_{i<j} X_{A \cap A_i \cap A_j} - \ldots + (-1)^n X_{A \cap \cap_{i=1}^n A_i}.$$

We shall always assume that our stochastic processes are additive. We note that a process with an (almost sure) additive extension to $\mathcal{C}$ also has an (almost sure) additive extension to $\mathcal{C}(u)$. Even without SHAPE, most of our examples will be seen to have additive sample paths, and we shall suppress the 'almost sure' in references to additivity unless some ambiguity arises.

We remark that $\mathcal{A}(u) \subseteq \mathcal{C}(u)$, and that for $B \in \mathcal{A}(u), X_B$ is

$\mathcal{F}_B$-measurable if X is adapted. However, in general for $C \in \mathcal{C}$, X_C is neither $\mathcal{G}_C$- nor $\mathcal{G}_C^*$-measurable. We do have the following:

Proposition 1.4.6 *Let $\mathcal{C}^\ell(\mathcal{A}')$ be the class of left-neighbourhoods of a finite sub-semilattice $\mathcal{A}'$ of $\mathcal{A}$. If X is adapted and $C_1, C_2 \in \mathcal{C}^\ell(\mathcal{A}')$, $C_1 \neq C_2$, then either X_{C_1} is $\mathcal{G}_{C_2}^*$-measurable or X_{C_2} is $\mathcal{G}_{C_1}^*$-measurable (or both).*

Proof. Let $\{A_1, A_2, ...\}$ be a numbering of $\mathcal{A}'$ consistent with the strong past. Then there exist i and j such that $C_1 = A_i \setminus \cup_{k=1}^{i-1} A_k$ and $C_2 = A_j \setminus \cup_{k=1}^{j-1} A_k$. If, say, $i < j$, then since $A_k \cap C_2 = \emptyset$ $\forall k \leq j-1$, it follows that $\mathcal{F}_{A_k} \subseteq \mathcal{G}_{C_2}^*, k = 1, ..., j-1$ and so X_{C_1} is $\mathcal{G}_{C_2}^*$-measurable. Note that in this case X_{C_2} may or may not be $\mathcal{G}_{C_1}^*$-measurable. □

Here we observe that the minimal outer-continuous filtration generated by an $\mathcal{A}$-indexed stochastic process can be defined as follows: for $A \in \mathcal{A}$,

$$\mathcal{F}_A = \cap_n \sigma(X_B, B \subseteq g_n(A), B \in \mathcal{A}).$$

In fact, it is easily seen that if $A \in \mathcal{A}(u)$ in the above definition, then $\mathcal{F}_A = \mathcal{F}_A^r$.

We now consider processes whose sample paths (trajectories) satisfy certain properties.

Definition 1.4.7 *(a) A function $f : \mathcal{A} \to \mathbf{R}$ is monotone outer-continuous if f is finitely additive on $\mathcal{C}$ and for any decreasing sequence (A_n) of sets in $\mathcal{A}(u)$ such that $\cap A_n \in \mathcal{A}(u)$, $f(\cap_n A_n) = \lim_n f(A_n)$.*
(b) A function $f : \mathcal{A} \to \mathbf{R}$ is monotone outer-continuous on $\mathcal{A}$ if f is finitely additive on $\mathcal{C}$ and for any decreasing sequence (A_n) of sets in $\mathcal{A}$, $f(\cap_n A_n) = \lim_n f(A_n)$.
(c) A function $f : \mathcal{A} \to \mathbf{R}$ is increasing if f is finitely additive on $\mathcal{C}$, $f(\emptyset) = 0$ and $f(C) \geq 0$, $\forall C \in \mathcal{C}$.

Monotone outer-continuous increasing functions may be regarded as distribution functions of measures on $\mathcal{B}$. This is a result of the next proposition which shows that such a function is continuous from above at the empty set on the algebra $\mathcal{C}(u)$, which generates $\mathcal{B}$.

Proposition 1.4.8 *(Dozzi, Ivanoff and Merzbach (1994)) Let $f : \mathcal{A} \to \mathbf{R}$ be a monotone outer-continuous increasing function. Then for any decreasing sequence (R_n) of sets in $\mathcal{C}(u)$ such that $\cap_n R_n = \emptyset$, we have $\lim_{n \to \infty} f(R_n) = 0$.*

THE PROBABILITY STRUCTURE

Proof. Note that by additivity and monotone outer-continuity, for any $\epsilon > 0$ and any $A \in \mathcal{A}(u)$, $0 \leq f(g_k(A) \setminus A) < \epsilon$ for every k sufficiently large. Thus, for any $C \in \mathcal{C}$ there exists $G \in \mathcal{C}$ such that $\overline{G} \subseteq C$ and $f(C \setminus G) < \epsilon$.

Since (R_n) is decreasing we have a compact set B_m as defined in Definition 1.1.1 #1 such that $R_n \subseteq B_m \; \forall n$. Now fix $\epsilon > 0$ and suppose that $R_n = \cup_{i=1}^{k(n)} C_i^n$, $C_i^n \in \mathcal{C}$. Choose G_i^n as above so that $G_i^n \in \mathcal{C}$, $\overline{G_i^n} \subseteq C_i^n$ and $f(E_i^n) < \epsilon/2^n k(n)$, where $E_i^n = C_i^n \setminus G_i^n$. Denote: $G_n = \cup_{i=1}^{k(n)} G_i^n$, $\overline{G_n} = \cup_{i=1}^{k(n)} \overline{G_i^n}$, $D_n = \cap_{k=1}^n G_k$, $\overline{D_n} = \cap_{k=1}^n \overline{G_k}$ and $S_n = \cup_{i=1}^{k(n)} E_i^n$. We have $R_n \subseteq D_n \cup \cup_{k=1}^n S_k$ and so $f(R_n) \leq f(D_n) + \sum_{k=1}^\infty f(S_k) < f(D_n) + \epsilon$. But $\overline{D_n} \subseteq R_n \searrow \emptyset$, and so by compactness, $\overline{D_n} = \emptyset$ for all n sufficiently large. Since ϵ is arbitrary, it follows that $\lim f(R_n) = 0$. □

Although additivity of a process was not defined as a sample path property, we note that additivity of each trajectory is implicit in the definitions of monotone outer-continuous and increasing processes which follow.

Definition 1.4.9 *(a) Let $X = \{X_A : A \in \mathcal{A}\}$ be a process with $X_\emptyset = X_{\emptyset'} = 0$. X is monotone outer-continuous if for each $\omega \in \Omega$, the function $X_\omega = X.(\omega) : \mathcal{A} \to \mathbf{R}$ is monotone outer-continuous. X is L^p-monotone outer-continuous if for any decreasing sequence (A_n) of sets in $\mathcal{A}(u)$ with $\cap_n A_n \in \mathcal{A}(u)$, $(E[|X_{A_n} - X_{\cap_n A_n}|^p])^{1/p} \to 0$ as $n \to \infty$.*
(b) Let $X = \{X_A : A \in \mathcal{A}\}$ be a process with $X_\emptyset = X_{\emptyset'} = 0$. X is monotone outer-continuous on $\mathcal{A}$ if for each $\omega \in \Omega$, the function $X_\omega = X.(\omega) : \mathcal{A} \to \mathbf{R}$ is monotone outer-continuous on $\mathcal{A}$. X is L^p-monotone outer-continuous on $\mathcal{A}$ if, for any decreasing sequence (A_n) of sets in $\mathcal{A}$, $(E[|X_{A_n} - X_{\cap_n A_n}|^p])^{1/p} \to 0$ as $n \to \infty$.

Definition 1.4.10 *Let $X = \{X_A : A \in \mathcal{A}\}$ be an integrable process with $X_\emptyset = X_{\emptyset'} = 0$. X is an increasing process if for each $\omega \in \Omega$, the function $X_\omega = X.(\omega) : \mathcal{A} \to \mathbf{R}$ is monotone outer-continuous and increasing.*

It is not always required that an increasing process be adapted. By uniform integrability, increasing processes are also L^1-monotone outer-continuous (that is, for any decreasing sequence (A_n) of sets in $\mathcal{A}(u)$ with $\cap_n A_n \in \mathcal{A}(u)$, $X_{A_n} \to_{L^1} X_{\cap A_n}$). We have the following corollary to Proposition 1.4.8.

Corollary 1.4.11 *Let X be an increasing process. Then for each*

$\omega \in \Omega$ fixed, $X(\omega)$ defines a unique σ-additive and σ-finite measure on $\mathcal{B}$.

1.5 Stopping Sets

The concept of stopping is central to martingale theory. It allows a characterization of the predictable σ-algebra, it leads to optional stopping theorems and characterizations of martingales and permits the development of a theory of local martingales. All of these problems will be thoroughly explored.

The main issue is the 'correct' definition of a stopping set, as many attempts to extend stopping time theory to a partially ordered index set have met with limited success. The following definition is analogous to that of Rozanov (1982), who called these sets 'co-compatible' sets.

Definition 1.5.1 *Let* $\xi : \Omega \to \mathcal{A}(u)$ *be of the form* $\xi(\omega) = \cup_{i=1}^{k} \xi_i(\omega), \xi_i : \Omega \to \mathcal{A}, i = 1, \ldots, k; k < \infty$. ξ *is called a stopping set (s.s.) if, for any* $A \in \mathcal{A}$, $\{\omega : A \subseteq \xi(\omega)\} \in \mathcal{F}_A$, $\{\omega : \phi = \xi(\omega)\} \in \mathcal{F}_\phi$, *and there exists a set* $B \in \mathcal{A}(u)$ *such that* $\xi \subseteq B$ *a.s. A stopping set* ξ *is called* simple *if* $\xi(\omega) \in \mathcal{A}$ $\forall \omega$. *(We write s.s.s. for a simple stopping set.)*

Comments:
1. Our definition appears opposite to that of Kurtz (1980a) and Hurzeler (1985b), who require measurability of the opposite inclusion $\{\omega : \xi(\omega) \subseteq A\}$. In fact, our definition is stronger, as will be seen in Lemma 1.5.8 below. We impose this stronger definition to ensure that the class of stopping sets is closed under finite intersections, a fact which will be required in the next chapter, and also for some of the optional sampling theorems in Chapter 3.
2. The assumption that $\xi \subseteq B$ a.s. is a boundedness assumption, which is trivially satisfied if $T \in \mathcal{A}(u)$.

The next three lemmas are from Ivanoff, Merzbach and Schiopu-Kratina (1993).

Lemma 1.5.2 *Let* ξ *and* ξ' *be two (simple) stopping sets. Then* $\xi \cap \xi'$ *defined by* $(\xi \cap \xi')(\omega) = \xi(\omega) \cap \xi'(\omega)$ *is a (simple) stopping set.*

Proof. Let $A \in \mathcal{A}$. Then $\{A \subseteq \xi \cap \xi'\} = \{A \subseteq \xi\} \cap \{A \subseteq \xi'\} \in \mathcal{F}_A$. Finally, $\{\emptyset = \xi \cap \xi'\} = \{\emptyset = \xi\} \cup \{\emptyset = \xi'\} \in \mathcal{F}_\emptyset$. □

STOPPING SETS

Lemma 1.5.3 *Assume that SHAPE holds. Let ξ and ξ' be two stopping sets. Then $\xi \cup \xi'$ defined by $(\xi \cup \xi')(\omega) = \xi(\omega) \cup \xi'(\omega)$ is a stopping set.*

Proof. Let $A \in \mathcal{A}$. Then by SHAPE $\{A \subseteq \xi \cup \xi'\} = \{A \subseteq \xi\} \cup \{A \subseteq \xi'\} \in \mathcal{F}_A$. Finally, $\{\emptyset = \xi \cup \xi'\} = \{\emptyset = \xi'\} \cap \{\emptyset = \xi'\} \in \mathcal{F}_\emptyset$. □

Comment: It follows from Lemma 1.5.3 that any finite union of simple stopping sets is a stopping set; however, the converse is not true: in general, a stopping set cannot be decomposed into a finite union of simple stopping sets.

We next will show that if SHAPE is satisfied, then any stopping set may be approximated from above by *discrete* stopping sets (i.e. stopping sets which take on only finitely many configurations).

Lemma 1.5.4 *Let ξ be a stopping set. If SHAPE holds or if ξ is simple and $g_n : \mathcal{A} \to \mathcal{A}_n \cup T$, then $g_n(\xi)$ is also a stopping set.*

Proof. We shall assume that SHAPE holds, as the proof for simple stopping sets is easier. First, observe that $\{\emptyset = g_n(\xi)\} = \{\emptyset = \xi\} \in \mathcal{F}_\emptyset$. Now it must be shown that $\{A \subseteq g_n(\xi)\} \in \mathcal{F}_A \ \forall A \in \mathcal{A}$. For $A \in \mathcal{A}$, define

$$d_n(A) = \bigcap_{\substack{B \in \mathcal{A} \\ A \subseteq g_n(B)}} B.$$

Note that both d_n and g_n are monotonic and that $d_n(g_n(A)) \subseteq A$ (by definition) and $A \subseteq g_n(d_n(A))$ (since g_n preserves arbitrary intersections).

Now, if $\xi = \cup_{i=1}^k \xi_i$ where for each i, $\xi_i \in \mathcal{A}$, then by SHAPE $\{A \subseteq g_n(\xi)\} = \cup_{i=1}^k \{A \subseteq g_n(\xi_i)\}$. Next,

$$\{A \subseteq g_n(\xi_i)\} \subseteq \{d_n(A) \subseteq d_n(g_n(\xi_i))\} \subseteq \{d_n(A) \subseteq \xi_i\}$$
$$\subseteq \{g_n(d_n(A)) \subseteq g_n(\xi_i)\} \subseteq \{A \subseteq g_n(\xi_i)\},$$

so $\{A \subseteq g_n(\xi_i)\} = \{d_n(A) \subseteq \xi_i\}$. Therefore, $\{A \subseteq g_n(\xi)\} = \{d_n(A) \subseteq \xi\} \in \mathcal{F}_{d_n(A)}$. Since $d_n(A) \subseteq A$, we are done. □

As SHAPE can be fairly restrictive, when T possesses a metric we can extend the previous lemma to more general indexing collections, including those of all of our examples. First we need the following:

Lemma 1.5.5 *If the condition of Lemma 1.3.4 holds and ξ is a stopping set then $\{B \subseteq \xi\} \in \mathcal{F}_B \ \forall B \in \tilde{\mathcal{A}}(u)$, where $\mathcal{F}_B$ is either $\mathcal{F}_B^0$ or $\mathcal{F}_B^r$.*

Proof. By Lemma 1.3.4, there exists a sequence (A_n) of sets in $\mathcal{A}$ such that $B = \overline{\cup_n A_n}$. Therefore, $\{B \subseteq \xi\} = \{\overline{\cup_n A_n} \subseteq \xi\} = \{\cup_n A_n \subseteq \xi\} = \cap_n \{A_n \subseteq \xi\} \in \vee_{n=1}^{\infty} \mathcal{F}_{A_n} \subseteq \mathcal{F}_B^0 \subseteq \mathcal{F}_B^r$. □

We are now in a position to prove the following.

Lemma 1.5.6 *If the condition of Lemma 1.3.4 holds and ξ is a stopping set, then $g_n(\xi)$ is also a stopping set.*

Proof. As before, $\{\emptyset = g_n(\xi)\} = \{\emptyset = \xi\} \in \mathcal{F}_\emptyset$. For $A \in \mathcal{A}(u)$, define
$$d_n(A) = \cap \{B \in \tilde{\mathcal{A}}(u) : A \subseteq g_n(B)\}.$$
We remark that $d_n(A) \in \tilde{\mathcal{A}}(u)$ since $\tilde{\mathcal{A}}(u)$ is closed under arbitrary intersections. As before, both d_n and g_n are monotonic, $d_n(g_n(A)) \subseteq A$ (by definition) and $A \subseteq g_n(d_n(A))$ (since by Lemma 1.1.3(b), g_n preserves arbitrary intersections in $\tilde{\mathcal{A}}(u)$). Now we may use the preceding lemma and proceed exactly as in the proof of Lemma 1.5.4 to show that $\{A \subseteq g_n(\xi)\} = \{d_n(A) \subseteq \xi\} \in \mathcal{F}_{d_n(A)}^0$. Since $d_n(A) \subseteq A$, $\mathcal{F}_{d_n(A)}^0 \subseteq \mathcal{F}_A$ and the proof is complete. □

The following lemma is an immediate consequence of Lemma 1.1.2 which implies that $\xi = \cap_n g_n(\xi)$, and Lemma 1.5.4 or Lemma 1.5.6 above.

Lemma 1.5.7 *If the conditions of either Lemma 1.5.4 or Lemma 1.5.6 are satisfied, then each stopping set may be approximated from above by a decreasing sequence of stopping sets, each taking finitely on many configurations.*

Lemma 1.5.7 now permits us to show that our definition of a stopping set is stronger than that of Kurtz (1980a).

Lemma 1.5.8 *(Ivanoff and Merzbach (1995)) Assume that the conditions of either Lemma 1.5.4 or Lemma 1.5.6 are satisfied. Then if ξ is a s.s., then $\{\xi = B\} \in \mathcal{F}_B^r$ and $\{\xi \subseteq B\} \in \mathcal{F}_B^r, \forall B \in \mathcal{A}(u)$.*

Proof. We begin by assuming that ξ is a discrete s.s., taking its values in $\mathcal{A}_p(u)$ for some p. Then if $B \in \mathcal{A}_p(u)$ and if $m \geq p$,

$$\begin{aligned}
\{\xi = B\} &= \{B \subseteq \xi\} \cap \cap_{\substack{A \in \mathcal{A}_p \\ A \not\subseteq B}} \{A \subseteq \xi\}^c \\
&= \{B \subseteq \xi\} \cap \cap_{\substack{A \in \mathcal{A}_p \\ A \not\subseteq B}} \cap_{n \geq m} \{(A \cap g_n(B)) \subseteq \xi\}^c \\
&\in \mathcal{F}_{g_m(B)}^0.
\end{aligned}$$

Thus, $\{\xi = B\} \in \cap_{m \geq p} \mathcal{F}^0_{g_m(B)} = \mathcal{F}^r_B$.

Then, for ξ a general stopping set, we have that $\xi = \cap_{n \geq m} g_n(\xi)$ $\forall m$, and so

$$\{\xi = B\} = \cap_{n \geq m}\{g_n(\xi) = g_n(B)\} \in \mathcal{F}^r_{g_m(B)}$$

$$\{\xi \subseteq B\} = \cap_{n \geq m}\{g_n(\xi) \subseteq g_n(B)\} \in \mathcal{F}^r_{g_m(B)}$$

for every m. The conclusion follows since $\mathcal{F}^r_B = \cap_m \mathcal{F}^r_{g_m(B)}$. □

When SHAPE is satisfied or if the conditions of Lemma 1.3.4 hold, Lemma 1.5.8 permits us to define the σ-algebra of events prior to a s.s. ξ in the natural way:

$$\mathcal{F}_\xi = \{F \in \mathcal{F} : F \cap \{\xi \subseteq B\} \in \mathcal{F}^r_B \quad \forall B \in \mathcal{A}(u)\}.$$

The inclusion in the definition may be replaced by equality if the stopping set is discrete. We note that if $\xi \equiv B$, then $\mathcal{F}_\xi = \mathcal{F}^r_B$. Recall that $\xi = \cap_n g_n(\xi)$. It is straightforward to verify that $\mathcal{F}_\xi = \cap_n \mathcal{F}_{g_n(\xi)}$ since $\mathcal{F}^r_B = \cap_n \mathcal{F}^0_{g_n(B)} = \cap_n \mathcal{F}^r_{g_n(B)}$ $\forall B \in \mathcal{A}(u)$. This observation leads to the following lemma from Ivanoff and Merzbach (1995):

Lemma 1.5.9 *Assume that the conditions of either Lemma 1.5.4 or Lemma 1.5.6 are satisfied. If ξ is a stopping set and if X is an adapted process which is monotone outer-continuous, then X_ξ is $\mathcal{F}_\xi$-measurable.*

Proof. First assume that ξ is a discrete stopping set. Then for any real number a and B a possible configuration of ξ,

$$\{X_\xi < a\} \cap \{\xi = B\} = \{X_B < a\} \cap \{\xi = B\} \in \mathcal{F}^r_B.$$

Thus, $\{X_\xi < a\} \in \mathcal{F}_\xi$, so X_ξ is $\mathcal{F}_\xi$-measurable.

For an arbitrary stopping set, by almost sure monotone outer-continuity $\{X_\xi < a\} = \cup_{M \geq m} \cap_{n \geq M} \{X_{g_n(\xi)} < a\} \cup Q$, where Q is a P-null set and m is arbitrary. Thus, X_ξ is $\cap_m \mathcal{F}_{g_m(\xi)} = \mathcal{F}_\xi$-measurable. □

For the remainder of this section, we shall assume that T is compact and that $\mathcal{F}_B = \mathcal{F}^r_B$.

Definition 1.5.10 *We define the σ-algebra of events strictly prior to the stopping set ξ, $\mathcal{F}_{\xi^-}$, to be the σ-algebra generated by $\mathcal{F}_{\emptyset'}$ and by all the sets of the form $F \cap \{B \subseteq \xi^\circ\}$ for any $B \in \mathcal{A}(u)$ and $F \in \mathcal{F}_B$.*

Proposition 1.5.11 *Let ξ, η be two stopping sets and assume that the conditions of Lemma 1.3.4 hold. Then:*
   (i) $\mathcal{F}_{\xi^-} \subseteq \mathcal{F}_\xi$ and if for all $A \in \mathcal{A}, \cup_n d_n(A) = A^\circ$ ($d_n(A) = \cap_{B \in \tilde{\mathcal{A}}(u),\ A \subseteq g_n(B)} B$), then $\{A \subseteq \xi\} \in \mathcal{F}_{\xi^-}\ \forall A \in \mathcal{A}$.
   (ii) If $\xi \subseteq \eta$ a.s., then $\mathcal{F}_{\xi^-} \subseteq \mathcal{F}_{\eta^-}$.
   (iii) If $F \in \mathcal{F}_\xi$ then $F \cap \{\xi \subseteq \eta^\circ\} \in \mathcal{F}_{\eta^-}$. In particular, $\{\xi \subseteq \eta^\circ\} \in \mathcal{F}_{\eta^-}$.
   (iv) If $\xi \subseteq \eta^\circ$ a.s. then $\mathcal{F}_\xi \subseteq \mathcal{F}_{\eta^-}$.

Proof. (i) We shall prove first that $\{A \subseteq \xi^\circ\} \in \mathcal{F}_A$, for any $A \in \mathcal{A}(u)$. Notice that $\{A \subseteq \xi^\circ\} = \cup_{n=1}^\infty \{g_n(A) \subseteq \xi\}$. The sequence of events $\{\omega : g_n(A) \subseteq \xi\}_n$ is increasing and the sequence of σ-algebras $\mathcal{F}_{g_n(A)}$ is decreasing, so $\cup_{k \geq n}\{g_k(A) \subseteq \xi\} \in \mathcal{F}_{g_n(A)}$, for every $n \geq 1$. It follows that

$$\cup_{n=1}^\infty \{g_n(A) \subseteq \xi\} \in \cap_{n=1}^\infty \mathcal{F}_{g_n(A)} = \mathcal{F}_A,$$

by the outer-continuity of the filtration $\{\mathcal{F}_A, A \in \mathcal{A}\}$ and since $\cap_n g_n(A) = A$. Therefore it follows that $\{A \subseteq \xi^\circ\} \in \mathcal{F}_A$.

Now we shall prove that the generators of $\mathcal{F}_{\xi^-}$ belong to $\mathcal{F}_\xi$. It is evident for $\xi = \emptyset'$. Take now any $F \in \mathcal{F}_A, A \in \mathcal{A}(u)$ and $B \in \mathcal{A}(u)$. Then, if $B \subseteq A$, or $B \not\subseteq A$ and $A \not\subseteq B$, then $F \cap \{A \subseteq \xi^\circ\} \cap \{\xi \subseteq B\} = \emptyset$. If $A \subseteq B$, it follows that $F \cap \{A \subseteq \xi^\circ\} \cap \{\xi \subseteq B\} \in \mathcal{F}_B$. ($F \cap \{A \subseteq \xi^\circ\} \in \mathcal{F}_A \subseteq \mathcal{F}_B$, and $\{\xi \subseteq B\} \in \mathcal{F}_B$ (see Lemma 1.5.8). Therefore, $F \cap \{A \subseteq \xi^\circ\} \in \mathcal{F}_\xi$.

Finally, $\{A \subseteq \xi\} = \cap_{n \geq 1}\{d_n(A) \subseteq \xi^\circ\} \in \mathcal{F}_{\xi^-}$.
(ii) Take any generator of $\mathcal{F}_{\xi^-}$, $F \cap \{A \subseteq \xi^\circ\}, F \in \mathcal{F}_A, A \in \mathcal{A}(u)$. Then

$$F \cap \{A \subseteq \xi^\circ\} = F \cap \{A \subseteq \xi^\circ\} \cap \{A \subseteq \eta^\circ\} \in \mathcal{F}_{\eta^-},$$

since $F \cap \{A \subseteq \xi^\circ\} \in \mathcal{F}_A$ and $\xi \subseteq \eta$. Therefore $\mathcal{F}_{\eta^-}$ contains the generators of $\mathcal{F}_{\xi^-}$.
(iii) Suppose first that ξ is a discrete s.s. and let B be any configuration of ξ. If $F \in \mathcal{F}_\xi$, then $F \cap \{\xi = B\} \in \mathcal{F}_B$, and $F \cap \{\xi \subseteq \eta^\circ\} \cap \{\xi = B\} = F \cap \{\xi = B\} \cap \{B \subseteq \eta^\circ\} \in \mathcal{F}_{\eta^-}$. Therefore, $F \cap \{\xi \subseteq \eta^\circ\} = \cup_{B \in \mathcal{H}} F \cap \{\xi = B\} \cap \{B \subseteq \eta^\circ\} \in \mathcal{F}_{\eta^-}$, where $\mathcal{H}$ is the set of all possible configurations of ξ.

Let ξ be a general s.s. Then $\xi = \cap_n g_n(\xi)$ and let $F \in \mathcal{F}_\xi = \cap_n \mathcal{F}_{g_n(\xi)}$. Since $\{\xi \subseteq \eta^\circ\} = \cup_n \{g_n(\xi) \subseteq \eta^\circ\}$, it follows that $F \cap \{\xi \subseteq \eta^\circ\} = \cup_{n \geq 1}\{F \cap \{g_n(\xi) \subseteq \eta^\circ\}\} \in \mathcal{F}_{\eta^-}$.
(iv) If $F \in \mathcal{F}_\xi$, then $F = F \cap \{\xi \subseteq \eta^\circ\} \in \mathcal{F}_{\eta^-}$ (by (iii)). □

STOPPING SETS 33

Proposition 1.5.12 *Let $\{\xi_n\}$ be a monotone sequence of stopping sets and assume the conditions of Proposition 1.5.11.*
(i) If ξ_n is decreasing and $\cap_n \xi_n = \xi$ is a stopping set, then $\cap_{n=1}^{\infty} \mathcal{F}_{\xi_n} = \mathcal{F}_\xi$.
(ii) If ξ_n is increasing and $\overline{\cup_n \xi_n} = \xi$ is a stopping set, then $\mathcal{F}_{\xi^-} = \vee_n \mathcal{F}_{\xi_n^-}$.

Proof. (i) $\xi \subseteq \xi_n$ for every n implies that $\mathcal{F}_\xi \subseteq \cap_{n=1}^{\infty} \mathcal{F}_{\xi_n}$. If $F \in \cap_n \mathcal{F}_{\xi_n}$, then

$$\begin{aligned} F \cap \{\xi \subseteq B\} &= F \cap (\cap_{m \geq M} (\cup_n (\xi_n \subseteq g_m(B))) \\ &= \cap_{m \geq M} [\cup_n (F \cap (\xi_n \subseteq g_m(B)))] \in \mathcal{F}_{g_M(B)}, \end{aligned}$$

$\forall M \in \mathbf{N}$, and it it follows that $F \cap \{\xi \subseteq B\} \in \cap \mathcal{F}_{g_M(B)} = \mathcal{F}_B$ and therefore $F \in \mathcal{F}_\xi$.

(ii) By Proposition 1.5.11(ii) $\vee_n \mathcal{F}_{\xi_n^-} \subseteq \mathcal{F}_{\xi^-}$. Consider any generator of $\mathcal{F}_{\xi^-}$: $F \cap \{B \subseteq \xi^\circ\}$, $F \in \mathcal{F}_B$. This set is equal to $F \cap \cup_n \{B \subseteq \xi_n^\circ\}$ and then is equal to $\cup_n \{F \cap \{B \subseteq \xi_n^\circ\}\} \in \vee_n \mathcal{F}_{\xi_n^-}$. □

The following corollary is a consequence of the previous two propositions.

Corollary 1.5.13 *Under the assumptions of Proposition 1.5.11,*
(i) If ξ_n is decreasing, $\cap_n \xi_n = \xi$ is a stopping set and $\xi \subseteq \xi_n^\circ$, for every n, then $\cap_{n=1}^{\infty} \mathcal{F}_{\xi_n^-} = \mathcal{F}_\xi$.
(ii) If ξ_n is increasing, $\overline{\cup_n \xi_n} = \xi$ is a stopping set and $\xi_n \subseteq \xi^\circ$, for every n, then $\mathcal{F}_{\xi^-} = \vee_n \mathcal{F}_{\xi_n}$.

CHAPTER 2

Predictability

The notion of predictability is fundamental to the theory of martingales and to stochastic integration, and provides the first step in the general theory of processes. Since the sets in $\mathcal{A}$ are only partially ordered, two different definitions of predictability arise. In this chapter we define two σ-algebras $\mathcal{P}$ and $\mathcal{P}^*$ of predictable sets in the product space $\Omega \times T$, using a given indexing collection $\mathcal{A}$ and a filtration $\{\mathcal{F}_A, A \in \mathcal{A}\}$. Several characterizations of the σ-algebra $\mathcal{P}$ via stopping sets will be proved. As will be shown, some of the characterizations obtained here hold even without requiring a topology on T. Announcable stopping sets will be defined and their relationship to $\mathcal{P}$ discussed. Progressivity and left-continuous processes are introduced in the last two sections of the chapter. Although of independent interest, neither section is required for the sequel. Much of the material presented in this chapter can be found in Ivanoff, Merzbach and Schiopu-Kratina (1993) and (1995).

Throughout this chapter, we assume that $T \in \mathcal{A}(u)$.

2.1 A characterization by stochastic intervals

Definition 2.1.1 *The* predictable σ-algebra $\mathcal{P}$ *is defined to be the σ-algebra generated by the sets of the form $F \times C$, where $C \in \mathcal{C}$ and $F \in \mathcal{G}_C$, in the product space $(\Omega \times T, \mathcal{F} \otimes \sigma(\mathcal{C}))$. If, instead of $\mathcal{G}_C$, we use $\mathcal{G}_C^*$, we obtain the *-predictable σ-algebra $\mathcal{P}^*$, also called the* weak predictable σ-algebra.

Denote by $\mathcal{P}_0$ ($\mathcal{P}_0^*$) the class of 'predictable rectangles' (' *- predictable rectangles') $F \times C, C \in \mathcal{C}$ and $F \in \mathcal{G}_C$ ($F \in \mathcal{G}_C^*$).

Remarks:
1. A set $F \times A$, with $A \in \mathcal{A}$ is (*-) predictable if $F \in \mathcal{F}_\emptyset$, since $\mathcal{G}_A = \mathcal{G}_A^* = \mathcal{F}_\emptyset$.
2. In the case $T = \mathbf{R}_+$, these definitions become the classical ones and $\mathcal{P} = \mathcal{P}^*$. In the two-parameter case, $\mathcal{P}$ and $\mathcal{P}^*$ are exactly

as defined in the literature on the two-parameter martingale theory (see for example Merzbach (1980) or Merzbach and Zakai (1980), in which more general predictable rectangles are also defined).
3. If T is a lattice, a predictable σ-algebra was defined by T. Norberg (1989b) for T-indexed filtrations. Another similar definition was suggested by M.F. Allain (1979).
4. Both definitions of predictability will be used extensively in the sequel for the Doob-Meyer decompositions and for the existence of quadratic variation.

With the exception of Proposition 2.1.8, all the results in this section are from Ivanoff, Merzbach and Schiopu-Kratina (1993).

Lemma 2.1.2 *The classes $\mathcal{P}_0$ and $\mathcal{P}_0^*$ are semi-algebras (i.e., the finite unions of sets from $\mathcal{P}_0$ (resp. $\mathcal{P}_0^*$) form an algebra).*

Proof. Denote by $r(\mathcal{P}_0)$ and $r(\mathcal{P}_0^*)$ the algebras generated by $\mathcal{P}_0$ and $\mathcal{P}_0^*$, respectively.

Let $F \times C \in \mathcal{P}_0$ and consider
$$(F \times C)^c = (F^c \times T) \cup (\Omega \times C^c) = ((F^c \times C) \cup (\Omega \times C^c)).$$
Since $F^c \in \mathcal{G}_C$, then $F^c \times C \in \mathcal{P}_0$. Let $C = A \setminus \cup_{i=1}^k A_i$. Then:
$$C^c = A^c \cup (\cup_{i=1}^k A_i) = (T \setminus A) \cup (\cup_{i=1}^k A_i) \in \mathcal{C}(u) = r(\mathcal{C}).$$
So, $\Omega \times C^c = (\Omega \times (T \setminus A)) \cup (\cup_{i=1}^k (\Omega \times A_i)) \in \mathcal{P}_0(u)$. Therefore $(F \times C)^c$ is a finite union of sets from $\mathcal{P}_0$.

Consider now the intersection: $(F_1 \times C_1) \cap (F_2 \times C_2) = (F_1 \cap F_2) \times (C_1 \cap C_2)$. Since $C_1 \cap C_2 \in \mathcal{C}$, it remains to show that $F_1 \cap F_2 \in \mathcal{G}_{C_1 \cap C_2}$. But this is clear since $\{\mathcal{G}_C, C \in \mathcal{C}\}$ is decreasing: $\mathcal{G}_{C_1} \cup \mathcal{G}_{C_2} \subseteq \mathcal{G}_{C_1 \cap C_2}$. Therefore $r(\mathcal{P}_0) = \mathcal{P}_0(u)$.

Exactly the same proof holds for $\mathcal{P}_0^*$. □

Comment. In what follows, we will find various characterizations of $\mathcal{P}$ via stopping sets. Although these characterizations cannot be extended to $\mathcal{P}^*$, it will be seen later that in fact it is the *-predictable σ-algebra which must be used to find martingale characterizations of processes, and subsequently to prove limit theorems for sequences of martingales.

Now we state and prove our first characterization of the predictable σ-algebra $\mathcal{P}$ by stochastic intervals. In this characterization, we discuss the connection between predictable sets and stopping sets. As will be shown, the topological structure is not

A CHARACTERIZATION BY STOCHASTIC INTERVALS

essential here and the existence of a natural partial order on $\mathcal{A}$ is enough in order to obtain this characterization.

Definition 2.1.3 *Let ξ and ξ' be two stopping sets. The ('inner-open' and 'outer-closed') stochastic interval $(\xi, \xi']$ is defined by*

$$(\xi, \xi'] = \{(\omega, t) : t \in \xi'(\omega) \setminus \xi(\omega)\}.$$

Other kinds of stochastic intervals will be defined later. Recall that a *discrete* stopping set is one which can take on only a finite number of configurations.

Lemma 2.1.4 *Assume SHAPE holds. The finite unions of stochastic intervals $(\xi, \xi']$ form an algebra. Moreover, the finite unions of stochastic intervals $(\xi, \xi']$ of discrete stopping sets also form an algebra.*

Proof. We have:

$$(\xi, \xi']^c = (\xi', T] \cup (\emptyset, \xi]$$

and for the intersection, we have:

$$(\xi_1, \xi_1'] \cap (\xi_2, \xi_2'] = (\xi_1 \cup \xi_2, \xi_1' \cap \xi_2'].$$

Using Lemmas 1.5.2 and 1.5.3, this completes the proof. □

Lemma 2.1.5 *If ξ is a discrete stopping set, then there exist finitely many disjoint sets $C_1, \ldots, C_k \in \mathcal{C}$, such that $(\emptyset, \xi]$ admits the representation:*

$$(\emptyset, \xi] = \bigcup_{i=1}^{k} F_i \times C_i,$$

where $F_i = \{\omega : C_i \subseteq \xi(\omega)\} \in \mathcal{G}_{C_i}, i = 1, \ldots, k.$

Proof. Suppose that $\xi(\omega) = \bigcup_{i=1}^{k} \xi_i(\omega)$, where $\xi_i : \Omega \to \mathcal{A}, i = 1, \ldots, k$.

There exists a partition $F_0, \ldots, F_n$ of Ω and sets (not necessarily disjoint) $A_i^j \in \mathcal{A}, i = 1, \ldots, k, j = 1, \ldots, n$ such that $\omega \in F_0 \Rightarrow \xi(\omega) = \emptyset$ and if $j \geq 1$, $\omega \in F_j \Rightarrow \xi_i(\omega) = A_i^j$ and $\xi(\omega) = \cup_i A_i^j$. Let the (finite) class $\mathcal{A}'$ denote the empty set and all possible intersections of one or more of the sets $A_i^j, j = 1, \ldots, n, i = 1, \ldots, k$. For an arbitrary non-empty set $B \in \mathcal{A}'$, recall that the left-neighbourhood of B in $\mathcal{A}'$ is $G(B) = B \setminus \cup_{B' \in \mathcal{A}', B \not\subseteq B'} B'$.

We note that if $B \not\subseteq \xi(\omega)$, then $G(B) \cap \xi(\omega) = \emptyset$. (This is clear by the definition of $G(B)$). From this observation, it follows that for any $A \in \mathcal{A}$ such that $A \cap G(B) \neq \emptyset$ and $A \subseteq B$, $\{\omega : B \subseteq \xi(\omega)\} = \{\omega : G(B) \subseteq \xi(\omega)\} = \{\omega : A \subseteq \xi(\omega)\}$. Let $F(B) = \{\omega : B \subseteq \xi(\omega)\}$.

Then $F(B) = \{\omega : G(B) \subseteq \xi(\omega)\} = \{\omega : A \subseteq \xi(\omega)\}$ for any $A \in \mathcal{A}$ such that $A \subseteq B$ and $A \cap G(B) \neq \emptyset$.

Now, $\xi = \cup_{i=1}^{k}\{(\omega,t) : t \in \xi_i(\omega)\} = \cup_{B \in \mathcal{A}'} F(B) \times B = \cup_{B \in \mathcal{A}'} F(B) \times G(B)$. Since $G(B) \in \mathcal{C}$, it remains to show that $F(B) \in \mathcal{G}_{G(B)}$. But for any set A such that $A \subseteq B$ and $A \cap G(B) \neq \emptyset$, $F(B) = \{\omega : A \subseteq \xi(\omega)\} \in \mathcal{F}_A$, since ξ is a stopping set. Thus $F(B) \in \cap_{A: A \cap G(B) \neq \emptyset} \mathcal{F}_A = \mathcal{G}_{G(B)}$. This completes the proof. $\square$

Comment: We note that the preceding lemma is true regardless of whether or not $T \in \mathcal{A}(u)$.

Proposition 2.1.6 *Assume SHAPE. Then $r(\mathcal{P}_0)$ is equal to the algebra generated by the stochastic intervals of the form $(\xi, \xi']$, where ξ and ξ' are discrete stopping sets.*

Proof. First consider $F \times C$, with $C = A \setminus \cup_{i=1}^{k} A_i$, and $F \in \mathcal{G}_C$. To avoid trivialities, assume $A \not\subseteq \cup_{i=1}^{k} A_i$. Then, without loss of generality, we assume $A_i \subseteq A, i = 1, \ldots, k$.

Define
$$\xi(\omega) = \begin{cases} A & \text{if } \omega \in F \\ \cup_{i=1}^{k} A_i & \text{if } \omega \notin F \end{cases}$$
and $\xi_i(\omega) = A_i$ (deterministic). Then $F \times C = \xi \setminus \cup_{i=1}^{k} \xi_i$.

Verify that ξ is a stopping set: For $B \in \mathcal{A}$,
$$\{\omega : B \subseteq \xi(\omega)\} = \begin{cases} \emptyset & \text{if } B \not\subseteq A \\ F & \text{if } B \subseteq A, B \not\subseteq \cup_{i=1}^{k} A_i \\ \Omega & \text{if } B \subseteq A, B \subseteq \cup_{i=1}^{k} A_i. \end{cases}$$

Therefore, in all cases, this set belongs to $\mathcal{F}_B$. Now:
$$\{\omega : \emptyset = \xi(\omega)\} = \begin{cases} \emptyset & \text{if } A \neq \emptyset \text{ and } \cup_{i=1}^{k} A_i \neq \emptyset \\ F^c & \text{if } A \neq \emptyset \text{ and } \cup_{i=1}^{k} A_i = \emptyset \\ \Omega & \text{if } A = \cup_{i=1}^{k} A_i = \emptyset. \end{cases}$$

If $\cup_{i=1}^{k} A_i = \emptyset$, then $\mathcal{G}_C = \mathcal{F}_\emptyset$ and therefore $\{\omega : \emptyset = \xi(\omega)\} \in \mathcal{F}_\emptyset$. The converse follows from Lemma 2.1.5. $\square$

Theorem 2.1.7 *Assume SHAPE. Then the σ-algebra of predictable sets $\mathcal{P}$ is generated by the stochastic intervals of the form $(\xi, \xi']$, where ξ and ξ' are stopping sets.*

Proof. This follows from Proposition 2.1.6 and Lemma 2.1.4. $\square$

Comment. There is no analogue to Theorem 2.1.7 for the σ-algebra $\mathcal{P}^*$.

A CHARACTERIZATION BY STOCHASTIC INTERVALS　　　　　39

Finally, consider a set $D \in r(\mathcal{P}_0)$; $D = \cup_{i=1}^{k} F_i \times C_i$, $C_i \in \mathcal{C}$, $F_i \in \mathcal{G}_{C_i}$, $i = 1, \ldots, k$. We define the début ξ_D of D as follows:

$$\xi_D(\omega) = \cup_{A \in \mathcal{A}_D(\omega)} A,$$

where $\mathcal{A}_D(\omega) = \{A \in \mathcal{A} : A \cap D(\omega) = \emptyset\}$ and $D(\omega) = \{t : (\omega, t) \in D\}$.

Proposition 2.1.8 *(Dozzi, Ivanoff and Merzbach (1994)) The début ξ_D of $D \in r(\mathcal{P}_0)$ is a discrete stopping set and $D \subseteq (\xi_D, T]$.*

Proof. Let $C_i = A_i \setminus \cup_{j=1}^{n_i} B_{ij} = A_i \setminus B_i$ be a maximal representation of $C_i, i = 1, \ldots, k$ in T, where $D = \cup_{i=1}^{k} F_i \times C_i$.

Let $\mathcal{E} = \{E_l; l = 1, \ldots, N\}$, where each set E_l is a finite intersection of the form $\cap B_{ij_i}$, where all the subscripts i must be distinct.

Let

$$\xi_l(\omega) = \begin{cases} E_l & \text{if } E_l \cap D(\omega) = \emptyset; \\ \emptyset & \text{if } E_l \cap D(\omega) \neq \emptyset. \end{cases}$$

We claim that $\xi_D(\omega) = \cup_{l=1}^{N} \xi_l(\omega)$.

To see this, note first that clearly, $\xi_l(\omega) \subseteq \xi_D(\omega)$, for all l. Conversely, $A \subseteq \xi_D(\omega) \Leftrightarrow A \cap D(\omega) = \emptyset \Leftrightarrow A \cap C_i = \emptyset$ for all i such that $\omega \in F_i$. But by the maximality of the representation of C_i, $A \subseteq B_i$ if $\omega \in F_i$. Let $E(\omega) = \cap_{i:\omega \in F_i} B_i$, and note that $A \subseteq E(\omega)$. Now, $D(\omega) = \cup_{i:\omega \in F_i} C_i$, and $B_i \subseteq C_i^c$, so $E(\omega) \subseteq \cap_{i:\omega \in F_i} C_i^c = (D(\omega))^c$. Thus, $E(\omega) \cap D(\omega) = \emptyset$, and so $A \subseteq E(\omega) \subseteq \cup_{l=1}^{N} \xi_l(\omega)$, completing the proof of the claim. Clearly, ξ_D takes only finitely many configurations.

It remains to show that $\{A \subseteq \xi_D\} \in \mathcal{F}_A$ for all $A \in \mathcal{A}$, and $\{\emptyset = \xi_D\} \in \mathcal{F}_\emptyset$. First,

$$\{\omega : \xi_D(\omega) = \emptyset\} = \{\omega : A \cap D(\omega) \neq \emptyset \; \forall \; A \in \mathcal{A}\} = \cup_{j:C_j \in \mathcal{A}} F_j.$$

But $C_j \in \mathcal{A} \Rightarrow \mathcal{G}_{C_j} = \mathcal{F}_\emptyset$, so $\{\omega : \xi_D(\omega) = \emptyset\} \in \mathcal{F}_\emptyset$.

Finally,

$$\{\omega : A \subseteq \xi_D(\omega)\} = \{\omega : A \cap D(\omega) = \emptyset\}$$

$$= \begin{cases} \Omega & \text{if } C_j \cap A = \emptyset \; \forall j = 1, \ldots, k \\ \cap_{j:C_j \cap A \neq \emptyset} F_j^c & \text{otherwise} \end{cases}$$

But $C_j \cap A \neq \emptyset \Rightarrow \mathcal{G}_{C_j} \subseteq \mathcal{F}_A \Rightarrow F_j^c \in \mathcal{F}_A$. Thus $\{\omega : A \subseteq \xi_D(\omega)\} \in \mathcal{F}_A$ for all $A \in \mathcal{A}$.

Thus ξ_D is a stopping set, and it is easily seen that $D \subseteq (\xi_D, T]$. □

2.2 Announcability

The concept of announcability is related to the possibility of foretelling a random event and therefore this idea is essential for the general theory. For example, an adapted and continuous process on $\mathbf{R}_+$ will be announcable since by continuity, the state or the knowledge at each time can be guessed by its past. In the classical theory, predictability and announcability are equivalent. However, in the general case, this equivalence does not hold, as was shown in the two-parameter case (cf. Merzbach (1979)). The goal of this section is to introduce announcable stopping sets and to characterize the predictable σ-algebra by using other kinds of stochastic intervals. The material presented in this section can be found in Ivanoff, Merzbach and Schiopu-Kratina (1993).

In this section, we will assume that the space T is a complete separable metric space.

Definition 2.2.1 *Given any two stopping sets ξ and ξ', we define the other stochastic intervals as follows:*

$$[\xi, \xi'] = \{(\omega, t) : t \in \xi'(\omega) \setminus (\xi(\omega))^\circ\}$$
$$[\xi, \xi') = \{(\omega, t) : t \in (\xi'(\omega))^\circ \setminus (\xi(\omega))^\circ\}$$
$$(\xi, \xi') = \{(\omega, t) : t \in (\xi'(\omega))^\circ \setminus \xi(\omega)\}$$

These sets can be called closed stochastic intervals, outer-open and inner-closed stochastic intervals, *and* open stochastic intervals, *respectively.*

Remark. For clopen sets, open and closed intervals are identical.

The definition of a 'inner-closed' stochastic interval leads naturally to the definition of an announcable stopping set:

Definition 2.2.2 *A stopping set ξ is called* announcable *if there exists an increasing sequence (ξ_n) of stopping sets such that for all $\omega \in \Omega$, $\cup_n \xi_n(\omega) = (\xi(\omega))^\circ$ on the set $\{\xi \neq \emptyset'\}$. Such a sequence (ξ_n) is called an* announcing sequence *for ξ.*

In order to thoroughly understand this definition and to point out that it is a generalization of a classical case, three comments are necessary.

Comments.

1. In the classical one-parameter case, the inclusion in the interior is replaced by the 'strictly-smaller' relation and this relation is required on the set where the stopping time is different from

zero. Here, we do not need this requirement on $\emptyset$ since $\emptyset^\circ = \emptyset$, but we need it on $\emptyset'$.

2. We define announcability with inclusion in the interior since we require that $[\xi, T] = \cap_n (\xi_n, T]$, if (ξ_n) announces ξ (see the proof of Theorem 2.2.5). This is true only if $\cup_n \xi_n = \xi^\circ$.

3. Notice that if ξ is clopen, we allow the announcing sequence (ξ_n) to be such that $(\xi_n) = \xi$. If there exist clopen sets other than $\emptyset$ and T, this does not give the classical definition of predictability. The reason for this is that, whenever the lattice is discrete, predictability (as in Definition 2.1.1) and adaptedness are the same. Predictability for such lattices must be defined in a different (but obvious) manner. We will not pursue this here.

We now make use of the fact that our space T is in fact a metric space (T, d), and make the following assumption (recalling that $A^\delta = \{t \in T : d(t, A) \leq \delta\}$):

Assumption 2.2.3 (Dilation)
Assume that if $A \in \mathcal{A}$, then $A^\delta \in \mathcal{A}$ $\forall \delta > 0$, that $(A \cap B)^\delta = A^\delta \cap B^\delta$ $\forall A, B \in \mathcal{A}, \forall \delta > 0$, and that $A^\circ \cup B^\circ = (A \cup B)^\circ, \forall A, B \in \mathcal{A}$. In addition, for each n there exists a dense subset D_n of $(0, 1]$ such that $\cup_{m > 1/\delta} A^{\delta - 1/m} = (A^\delta)^\circ$ for all $A \in \mathcal{A}_n$ and all $\delta \in D_n$.

Comments.

1. If the mapping $f_\delta : A \to A^\delta$ is restricted to sets $A \in \mathcal{A}_n$, then this assumption implies that f_δ has a lower adjoint, $d_\delta : A \to \mathcal{A}_n$. (See Gierz et al. (1980) for the definition of a lower adjoint.)

2. Note that $A = \cap_n A^{1/n} = \cap_n (A^{1/n})^\circ$.

3. Notice that this assumption holds when $T = \mathbf{R}_+^n$ and $\mathcal{A}$ is the lattice of rectangles $\{[0, x] : x \in \mathbf{R}_+^n\}$. In fact, in this case, $\cup_{n > 1/\delta} A^{\delta - 1/n} = (A^\delta)^\circ$ $\forall A \in \mathcal{A}$, $\forall \delta > 0$. However, this is not true if $T = [0, 1]^n$, but the dilation assumption is still satisfied. Consider the set $[0, 1 - d]^n$. Then if $\delta = d$, $\cup_{m > 1/d} A^{d - 1/m} = [0, 1)^n$ while $(A^d)^\circ = [0, 1]^n$.

Lemma 2.2.4 *Assume SHAPE and the dilation assumption. If $\xi = \cup_{i=1}^k \alpha_i$ is a stopping set, where $\alpha_i : \Omega \to \mathcal{A}_n, i = 1, \ldots, k$, then $\xi^\delta = \cup_{i=1}^k \alpha_i^\delta$ is an announcable stopping set, for all $\delta \in D_n$.*

Proof. By Comment 1 above, a proof analogous to that of Lemma 1.5.4 or Lemma 1.5.6 shows that ξ^δ is a stopping set.

Now, to show announcability, for $m > \delta^{-1}$ let us define $\xi_m = \xi^{\delta - 1/m} = \cup_{i=1}^k \alpha_i^{\delta - 1/m} = \cup_{i=1}^k \alpha_{i,m}$. This is an increasing sequence of stopping sets. If $\delta \in D_n$, by dilation, $\cup_m \alpha_{i,m} = (\alpha_i^\delta)^\circ$, and so also by dilation, $\cup_m \xi_m = \cup_{i=1}^k \cup_m \alpha_{i,m} = \cup_{i=1}^k (\alpha_i^\delta)^\circ = (\cup_{i=1}^k \alpha_i^\delta)^\circ = (\xi^\delta)^\circ$. Thus (ξ_m) announces ξ^δ if $\delta \in D_n$. □

We are now in a position to state and prove the main result of this paragraph.

Theorem 2.2.5 *Let T be a complete separable metric space, and $\mathcal{A}$ be an indexing collection of T, satisfying SHAPE and the dilation condition. The predictable σ-algebra $\mathcal{P}$ is generated by the following:*

1. $\{F \times C, C \in \mathcal{C}, F \in \mathcal{G}_C\}$.
2. *The stochastic intervals $(\xi, \xi']$, where ξ and ξ' are stopping sets.*
3. *The closed stochastic intervals $[\xi, \xi']$, where ξ and ξ' are stopping sets, and ξ is announcable.*
4. *The outer-open and inner-closed stochastic intervals $[\xi, \xi')$, where ξ and ξ' are both announcable stopping sets.*

Proof. The equivalence between 1. and 2. have already been proven under weaker assumptions. Let us show that the σ-algebras in 2. and 3. are equal.

First note that $[\xi, \xi'] = [\xi, T] \setminus (\xi', T]$ and suppose that ξ is announcable. If (ξ_n) is an announcing sequence for ξ, we have $[\xi, T] = \cap_n (\xi_n, T]$. Conversely, $(\xi, \xi'] = (\xi, T] \cap (\emptyset, \xi']$. Now suppose $\xi = \cup_{i=1}^k \alpha_i$ where $\alpha_i : \Omega \to \mathcal{A}_n, i = 1, \ldots, k$. Then $(\xi, T] = \cup_{\delta_m} [\xi^{\delta_m}, T]$, where (δ_m) is any decreasing sequence in D_n such that $\delta_m \to 0$ (see Comment 2 following Assumption 2.2.3), and ξ^{δ_m} is announcable $\forall m$ by Lemma 2.2.4. For general ξ, we note that Lemma 1.5.7 implies that ξ may be approximated from above by a decreasing sequence of stopping sets $(g_n(\xi))$, where $g_n(\xi) \in \mathcal{A}_n(u)$. Thus, $(\xi, T] = \cup_n (g_n(\xi), T]$, and 2. and 3. are the same.

The fact that the σ-algebras in 3. and 4. are the same follows by observing that $[\xi, \xi') = [\xi, \xi'] \setminus [\xi', \xi']$ and conversely, $[\xi, \xi'] = \cap_n [\xi, g_n(\xi')]$, by Lemma 1.5.7. Note that $g_n(\xi')$ is a finite union of sets taking values in $\mathcal{A}_n$. But $[\xi, g_n(\xi')] = \cap_{\delta_m} [\xi, (g_n(\xi'))^{\delta_m})$, where (δ_m) is any decreasing sequence in D_n such that $\delta_m \to 0$. As $(g_n(\xi'))^{\delta_m}$ is announcable by Lemma 2.2.4, this completes the proof that 3. and 4. are the same. □

Corollary 2.2.6 *If ξ is an announcable stopping set, then $\partial \xi = [\xi]$ is a predictable set.*

Proof. Let (ξ_n) be an announcing sequence for ξ. Then $\partial \xi = \cap_n (\xi_n, \xi]$. □

Remarks:

1. The most natural example is $T = \mathbf{R}_+^n$ and $\mathcal{A} = \{[0,t] : t \in \mathbf{R}_+^n\}$. For $n = 2$, the preceding characterization of $\mathcal{P}$ has already been given (Merzbach (1980) or Meyer (1981)). This example can be easily extended to the case $T = \mathbf{R}_+ \times \mathbf{R}^n$, or $T = \mathbf{R}^n$ with $n \leq \infty$, which appears in the theory of stochastic differential equations.

2. In the classical case, it follows from the predictable section theorem that the converse of the corollary holds: Any predictable stopping time τ (i.e., a stopping time whose graph $[\tau, \tau] = [\tau]$ belongs to $\mathcal{P}$) is announcable. However, already in the two-parameter case the situation is more delicate: the result holds for stopping lines in the plane (see Merzbach (1980) or Bakry (1981)), but not for stopping points. In fact, in our context this means that it is simpler to approximate a stopping set from below by stopping sets taking values in $\mathcal{A}(u)$, than to approximate a simple stopping set from below by simple stopping sets taking values in $\mathcal{A}$.

2.3 Progressivity

In this section, we introduce the concept of the progressive σ-algebra on the product space $\Omega \times T$; we study its relationships to random sets and finally to stopping sets and to the predictable σ-algebra.

The role of progressivity in the general theory is not really essential and therefore this section is not used in the sequel, and may be skipped at the first reading. Most of the material in this section can be found in Ivanoff, Merzbach and Schiopu-Kratina (1995).

We suppose in this section that T is a complete separable metric space with metric d. The class $\mathcal{A}$ may be very general, $\mathcal{A}$ need not be an indexing collection, and it is enough to require the existence of a countable dense subset $\{t_n\}$ of T such that for every $A \in \mathcal{A}$, $A = \overline{\cup_{t_n \in A} \{t_n\}}$. Note that this last assumption is satisfied if all sets in $\mathcal{A}$ are domains (A is a domain if $A \subseteq \overline{(A^\circ)}$).

Let us begin with some remarks on random sets:

Definition 2.3.1 $\xi : \Omega \to \mathcal{A}$ *is called a* simple random set *if* $\{\omega : \xi(\omega) \subseteq S\} \in \mathcal{F}$ *for all closed sets* S. ξ *is called a* random set *if there exist simple random sets* ξ_i, $i = 1, \ldots, k < \infty$ *such that* $\xi = \cup_{i=1}^k \xi_i$.

Lemma 2.3.2 *The following statements are equivalent, for $\xi : \Omega \to \mathcal{A}$.*

(i) ξ *is a simple random set.*

(ii) $\{\omega : \xi(\omega) \subseteq O\} \in \mathcal{F}$ *for all open sets O.*

(iii) $\{\omega : \xi(\omega) \cap S = \emptyset\} \in \mathcal{F}$ *for all closed sets S.*

(iv) $\{\omega : \xi(\omega) \cap O = \emptyset\} \in \mathcal{F}$ *for all open sets O.*

(v) $\{\omega : O \subseteq \xi(\omega)\} \in \mathcal{F}$ *for all open sets O.*

(vi) $\{\omega : S \subseteq \xi(\omega)\} \in \mathcal{F}$ *for all closed sets S.*

(vii) $\forall t \in T$, $d(\xi, t)$ *is a random variable (where $d(\xi,t)(\omega) = d(\xi(\omega), t)$).*

(viii) The map $\omega \to 1_{\xi(\omega)}(t)$ is $\mathcal{F}$-measurable.

Proof. For $\epsilon > 0$, denote $A^{-\epsilon} = \{t \in T : d(t, A^c) \geq \epsilon\}$. By convention, $T^{-\epsilon} = T$, $\forall \epsilon > 0$. It is clear that (iii) is equivalent to (ii), and (iv) to (i).

(i) $\Rightarrow$ (ii): Let O be an open set. Then
$$\{\xi \subseteq O\} = \cup_n \{\xi \subseteq O^{-1/n}\} \in \mathcal{F}.$$

(ii) $\Rightarrow$ (i): Let S be a closed set. Then
$$\{\xi \subseteq S\} = \cap_n \{\xi \subseteq (S^{1/n})^\circ\} \in \mathcal{F}.$$

(iv) $\Rightarrow$ (vii):
$$\{d(\xi, t) \leq \alpha\} = \cap_n \{\xi \cap B(t, \alpha + \frac{1}{n}) \neq \emptyset\},$$
where $B(t, \alpha)$ is the open ball with center t and radius α.

(vii) $\Rightarrow$ (iv): It is enough to suppose (vii) for the dense sequence $\{t_n\}$ in T. Indeed, the family $B(t_n, r)$, $r \in Q$ is an open base in T, and
$$\{\xi \cap B(t_n, r) = \emptyset\} = \{d(\xi, t_n) \geq r\} \in \mathcal{F}.$$
Therefore, for any open set O, we have
$$\{\xi \cap O \neq \emptyset\} = \bigcup_{B(t_n,r) \subseteq O} \bigcup_r \{\xi \cap B(t_n, r) \neq \emptyset\} \in \mathcal{F}.$$

(iii) $\Rightarrow$ (viii): For each t fixed, $\omega \to 1_{\xi(\omega)}(t) = 1_{\{\omega : t \in \xi(\omega)\}}$ and $\{\omega : t \in \xi(\omega)\} = \{\omega : \xi(\omega) \cap \{t\} \neq \emptyset\} \in \mathcal{F}$.

(viii) $\Rightarrow$ (iv): Let O be an open set and let $\{t_n\}$ be the dense sequence in T. Then $\{\omega : \xi(\omega) \cap O \neq \emptyset\} = \cup_{t_n \in O} \{\omega : t_n \in \xi(\omega)\} \in \mathcal{F}$.

Properties (v) and (vi) follow immediately from the other statements and from the fact that (viii) is equivalent to the measurability of the map $\omega \to 1 - 1_{\xi(\omega)}(t) = 1_{\xi(\omega)^c}(t)$. □

Remarks.
1. If we do not assume the existence of a dense sequence, property (viii) is not equivalent to the other properties as the following example shows: Let $\Omega = T = \mathbf{R}$, let $\mathcal{A}$ be all the finite sets in $\mathbf{R}$, and let $\mathcal{F}$ be the σ-algebra generated by the finite sets. Then the map $\omega \to \{\omega\}$ is not a random set although property (viii) is verified.
2. If $\mathcal{A}$ is the family of compact subsets of T, then $\xi : \Omega \to \mathcal{A}$ is a simple random set if and only if ξ is measurable with respect to the Borel σ-algebra $\mathcal{B}(\mathcal{A})$ generated by the Hausdorff topology on $\mathcal{A}$. That is, $\{\omega : \xi(\omega) \in \alpha\} = \xi^{-1}(\alpha) \in \mathcal{F}$, for any $\alpha \in \mathcal{B}(\mathcal{A})$ (Valadier (1971)).
3. From Lemma 2.3.2 (viii), it follows that any intersection of simple random sets is a simple random set, and if $\mathcal{A} = \mathcal{A}(u)$, then any finite union of random sets is still a random set.

Proposition 2.3.3 *Let $\xi : \Omega \to \mathcal{A}$ be such that ξ takes on finitely many configurations. Then ξ is a simple random set if and only if $(\emptyset, \xi] = \{(\omega, t) : t \in \xi(\omega)\} \in \mathcal{F} \times \mathcal{B}$.*

Proof. Since T is a Polish space, it follows directly by the capacitability theorem, just as in the classical case. □

For any set S, $\mathcal{F}_S$ can be defined to be $\vee_{A \in \mathcal{A}, A \subseteq S} \mathcal{F}_A$. Therefore, a simple random set ξ is a simple stopping set if and only if $\{\omega : S \subseteq \xi(\omega)\} \in \mathcal{F}_S$ for any set S which is the closure of a set of the form $\cup_{A \in \mathcal{A}, A \subseteq S} A$. Indeed, for such a set S, we have $\{\omega : S \subseteq \xi(\omega)\} = \cap_{A \in \mathcal{A}, A \subseteq S} \{\omega : A \subseteq \xi(\omega)\}$, and since we assume the existence of a countable dense subset of T, this intersection can be restricted to be countable. In the case where $\mathcal{A}$ is the class of all the closed domains of T, we have the following characterization which will show that the definition of a simple stopping set in the previous chapter is very natural, even if $\mathcal{A}$ is not an indexing collection.

Proposition 2.3.4 *Assume that $\mathcal{A}$ is the class of all closed domains of T. Let ξ be a simple random set. Then ξ is a simple stopping set if and only if for any closed set A, $d(A, \xi) = \inf\{d(a, x), a \in A, x \in \xi\}$ is a stopping time with respect to the filtration $\{\mathcal{F}^A_\alpha, \alpha \geq 0\}$, where $\mathcal{F}^A_\alpha = \mathcal{F}_{A^\alpha}$.*

Proof. Suppose ξ is a simple stopping set. Then for any nonempty open set O and domain S, $\{\omega : O \subseteq (\xi(\omega) \cap S)^\circ\} \in \mathcal{F}_S$ since

$\{\omega : O \subseteq (\xi(\omega) \cap S)^\circ\} = \{\omega : O \subseteq \xi(\omega) \cap S\} = \{\omega : \overline{O} \subseteq \xi(\omega)\} \cap \{\overline{O} \subseteq S\}$, and, therefore, $\{\omega : (\xi(\omega) \cap S)^\circ \neq \emptyset\} = \cup_i \{\omega : B_i \subseteq (\xi(\omega) \cap S)^\circ\}$, where $\{B_i\}$ is a countable open base of T. Thus, for any closed domain S, we have $\{\omega : (\xi(\omega) \cap S)^\circ \neq \emptyset\} \in \mathcal{F}_S$. For $\alpha \geq 0$, $\{\omega : d(A, \xi(\omega)) \leq \alpha\} = \cap_n \{\omega : (\xi(\omega) \cap A^{\alpha+1/n})^\circ \neq \emptyset\} \in \mathcal{F}_{A^\alpha}$, by the right continuity of the filtration $\{\mathcal{F}^A_\alpha, \alpha \geq 0\}$.

Conversely, consider singletons $\{t_n\}$, where $\{t_n\}$ is a dense sequence in T. Then, $\{\omega : d(t_n, \xi(\omega)) \leq \alpha\} = \{\omega : \xi(\omega) \cap B(t_n, \alpha) \neq \emptyset\} \in \mathcal{F}_{B(t_n, \alpha)}$, where $B(t_n, \alpha)$ is the closed ball of center t_n and radius α. These balls form a basis for the topology. Then for each closed domain $S \in \mathcal{A}$, we have that

$$\{\omega : S \subseteq \xi(\omega)\} = \cap_{m > M} \cap_{t_n \in S^\circ} \{\omega : \xi(\omega) \cap B(t_n, 1/m) \neq \emptyset\}$$

for any positive integer M. But by definition, if $m > M$ and $t_n \in S^\circ$, $\mathcal{F}_{B(t_n, 1/m)} \subseteq \mathcal{F}_{S^{1/m}}$, and so $\{\omega : S \subseteq \xi(\omega)\} \in \mathcal{F}_{S^{1/M}}$, $\forall M$. Hence, by right continuity, $\{\omega : S \subseteq \xi(\omega)\} \in \mathcal{F}_S$, and thus ξ is a simple stopping set. □

Comment. In view of Propositions 2.3.3 and 2.3.4, it would seem reasonable to require that any stopping set be a random set if T is a complete separable metric space. In fact, in most of the applications in which stopping sets have been defined, it is easily verified that they are random sets as well, when T is Polish.

Definition 2.3.5 *The σ-algebra of progressive sets Π is the σ-algebra of sets Γ in the product space $\mathcal{F} \times \mathcal{B}(T)$ such that for all $A \in \mathcal{A}$, $\Gamma \cap (\Omega \times A) \in \mathcal{F}_A \times \mathcal{B}(A)$, where $\mathcal{B}(A)$ is the class of Borel sets of T contained in A.*

Theorem 2.3.6 *Assume that $\mathcal{A}$ is an indexing collection, that $g_n : \mathcal{A} \to \mathcal{A}_n$ and that SHAPE holds. Let ξ be a random set. Then ξ is a stopping set if and only if the stochastic interval $(\emptyset, \xi]$ is a progressive set.*

Proof. Suppose that $\Gamma = (\emptyset, \xi]$ is a progressive set. Then for all $A \in \mathcal{A}$, $\Gamma^c \cap (\Omega \times A) \in \mathcal{F}_A \times \mathcal{B}(A)$. By the capacitability theorem, its projection on Ω is still measurable: i.e., $\text{Proj}(\Gamma^c \cap (\Omega \times A)) \in \mathcal{F}_A$. But this set is equal to $\{\omega : \exists t \in A : (\omega, t) \notin \Gamma\} = \{\omega : A \subseteq \xi(\omega)\}^c$. Therefore, ξ is a stopping set.

Conversely, let ξ be a stopping set. Suppose first that ξ is a simple stopping set taking its configurations in $\mathcal{A}_n$. Let $A \in \mathcal{A}_n$.

Then since $\xi \cap A \in \mathcal{A}_n$,

$$(\emptyset, \xi] \cap (\Omega \times A) = (\emptyset, \xi \cap A] = \bigcup_{\substack{B \subseteq A \\ B \in \mathcal{A}_n}} \{\omega : B \subseteq \xi(\omega)\} \times B \in \mathcal{F}_A \times \mathcal{B}(A).$$

Now, for ξ a simple stopping set and $A \in \mathcal{A}$,

$$\begin{aligned}
(\emptyset, \xi] \cap (\Omega \times A) &= (\emptyset, \xi \cap A] \\
&= \bigcap_{m \geq n} (\emptyset, g_m(\xi) \cap g_m(A)] \\
&\in \mathcal{F}_{g_n(A)} \times \mathcal{B}(g_n(A)) \; \forall n.
\end{aligned}$$

But since $A = \cap_{m \geq n} g_m(A)$, $\forall n$, and the filtration is monotone outer-continuous, it follows that $(\emptyset, \xi] \cap (\Omega \times A) \in \mathcal{F}_A \times \mathcal{B}(A)$.

For a general stopping set $\xi = \cup_{i=1}^{k} \alpha_i$, the proof proceeds exactly as above, assuming first that $\alpha_1, \ldots, \alpha_k$ take their values in $\mathcal{A}_n$, and then by approximating an arbitrary stopping set from above. □

Corollary 2.3.7 *Assume that $\mathcal{A}$ is an indexing collection, that $g_n : \mathcal{A} \to \mathcal{A}_n$ and that SHAPE holds. Then the σ-algebra of predictable sets is included in the σ-algebra of progressive sets.*

Proof. It follows from Theorem 2.3.6 that if ξ is a stopping set, then $(\emptyset, \xi]$ is a progressive set. Therefore, the result follows from Theorem 2.1.7 and the fact that: $(\xi, \xi'] = \{(\omega, t) : t \in \xi'(\omega) \setminus \xi(\omega)\} = (\emptyset, \xi'] \setminus (\emptyset, \xi]$. □

2.4 Left-continuous processes

In this last section of the chapter, we do not require a topology on T and we assume that the collection $\mathcal{A}$ is a distributive lattice of subsets of T closed under countable intersections and such that $\emptyset, T \in \mathcal{A}, A \wedge B = A \cap B, A \vee B \subseteq A \cup B$. We note that $A \leq B$ in the lattice if and only if $A \subseteq B$.

The goal of this section is to define 'left-continuous' functions $f : T \to \mathbf{R}$ without any topology on T, in such a way that the left-continuous processes adapted to the given filtration, characterize the predictable σ-algebra. The material here is also included in Ivanoff, Merzbach and Schiopu-Kratina (1993) and it will be shown that our characterization is also valid in Norberg's setting (1989b), where we may identify $\mathcal{A}$ with T.

It is important in what follows to distinguish between points of the space T and elements of the lattice $\mathcal{A}$. Processes in this section will be T-indexed, while the filtration is $\mathcal{A}$-indexed. For an increasing sequence $(A_n) \subseteq A$, we write $A_n \uparrow A$ if $\vee A_n$ exists and $\vee A_n = A$. We will need the following assumption on the lattice $\mathcal{A}$, which will replace separability from above as defined in Definition 1.1.1:

Assumption 2.4.1 (Lattice separability)
There exists an increasing sequence $(\mathcal{A}_n)$ of finite sublattices of $\mathcal{A}$ (with $\wedge_n = \wedge$ and $\vee_n = \vee$), each containing $\emptyset$ and T, such that, for every $A \in \mathcal{A}$, $A \neq \emptyset$, there exists $(A)_n \in \mathcal{A}_n$, such that $(A)_n \subset A$, $\forall n$ and $(A)_n \uparrow A$, if $A \neq \emptyset'$. Also, we assume that $(A)_n = ((A)^n)_n$, for every $n \geq 1$, where $(A)^n = \wedge_{A' \in \mathcal{A}_n, A' \supseteq A} A'$.

Comment. In a complete lattice, a natural way to define $(A)_n$ is by using the 'way below' relation '$\ll$' (see Gierz et al. (1980) for the definition), i.e., $(A)_n = \vee_{A' \in \mathcal{A}_n, A' \ll A} A'$, and require that $A = \sup(A)_n$ and $((A)^n)_n = (A)_n$. Note that $\mathcal{A}$ is necessarily a continuous lattice (see Definition 1.6, p. 41 of Gierz et al. (1980)).

Assumption 2.4.2 (TIP)
For all $A \in \mathcal{A}, A \neq \emptyset$, there exists a (not necessarily unique) point $t \in A$ such that $t \in A \setminus B$ for any $B \in \mathcal{A}$, such that $A \not\subseteq B$.

Remark. The axiom of choice allows us to choose one such point $t_A \in A$ for each $A \in \mathcal{A}, A \neq \emptyset$. As well, we define $t_\emptyset = t_{\emptyset'}$. We will fix such a collection $\{t_A : A \in \mathcal{A}\}$, and we will refer to its elements as 'tips' of the sets $A, A \in \mathcal{A}$.

As processes in this section are T-indexed while the filtration is $\mathcal{A}$-indexed, we must give a definition of $\mathcal{F}_\mathcal{A}$-adaptedness for the T-indexed processes:

Definition 2.4.3 $X : \Omega \times T \to \mathbf{R}$ *is $\mathcal{F}_\mathcal{A}$-adapted if $X(\cdot, t)$ is $\mathcal{F}_A$-measurable, $\forall A \in \mathcal{A}$ such that $t \in A$. If $t = t_\emptyset$, then $X(\cdot, t_\emptyset)$ is $\mathcal{F}_\emptyset$-measurable.*

This definition leads naturally to the concept of 'left-continuity' on T.

Definition 2.4.4 *Let $t, t_n \in T, n \geq 1$. We say that t_n converges to t from the left $(t_n \to t-)$ if $\forall A, A_i \in \mathcal{A}, i = 1, \ldots, k$ such that $t \in A \setminus \cup_{i=1}^k A_i \Rightarrow \exists n_0$ such that $t_n \in A \setminus \cup_{i=1}^k A_i, \forall n \geq n_0$.*

Comment. It is clear that the limit of a left-convergent sequence is *not* necessarily unique (see Example 2.4.7 below).

Definition 2.4.5 *A function $f : T \to \mathbf{R}$ is left-continuous if $t_n \to t- \Rightarrow f(t_n) \to f(t)$ in $\mathbf{R}$.*

LEFT-CONTINUOUS PROCESSES 49

Comment. If a function f is left-continuous, it must take the same value on all limits of each convergent sequence.

We now state the main theorem of this section.

Theorem 2.4.6 *Assume that $\mathcal{A}$ satisfies Assumptions 2.4.1 and 2.4.2. Then $\mathcal{P}$ is generated by the $\mathcal{F}_A$-adapted left-continuous processes on $\Omega \times T$.*

Proof. We begin by showing that I_P is left-continuous and adapted for any generator P of $\mathcal{P}$. Let $P = F \times C, C \in \mathcal{C}, F \in \mathcal{G}_C$. To show adaptedness of $I_{F \times C}$, note that if $t \notin C$, $I_{F \times C}(\cdot, t) = 0$. If $t \in C$, $I_{F \times C}(\cdot, t) = I_F(\cdot)$. We must therefore show that $F \in \mathcal{F}_A$, whenever $t \in A \cap C$. But $A \cap C \neq \emptyset \Rightarrow \mathcal{G}_C \subseteq \mathcal{F}_A$, so $F \in \mathcal{F}_A$, as required. If $P = F \times A', F \in \mathcal{F}_\emptyset$, $A' \in \mathcal{A}$, note that $\mathcal{F}_\emptyset \subseteq \mathcal{F}_A \ \forall A$, so I_P is adapted. The proof of left-continuity is straightfoward. It is clear from Definition 2.4.4 that if $t_n \to t-$, then $I_P(\omega, t_n) = I_P(\omega, t)$ for all n sufficiently large.

Conversely, we must show now that every adapted left-continuous process X is $\mathcal{P}$-measurable. Let $(\mathcal{A}_n)$ be a sequence of finite sublattices of $\mathcal{A}$ with the properties prescribed by Assumption 2.4.1. For each $t \in T$ let $A_t^n = \wedge_{A' \in \mathcal{A}_n, t \in A'} A'$ (i.e., A_t^n is the smallest set in $\mathcal{A}_n$ containing the point t).

For each $A \in \mathcal{A}_n$, let $G^n(A)$ denote the left-neighbourhood of A in $\mathcal{A}_n$. For each n, the sets $(G^n(A) : A \in \mathcal{A}_n)$ form a partition of T.

Given a left-continuous adapted process X, define X_n as follows:

$$\begin{aligned} X_n(\omega, t) &= \sum_{A \in \mathcal{A}_n} X(\omega, t_{(A)_n}) I_{G^n(A)}(t) \\ &= X(\omega, t_n), \text{ where } t_{(A_t^n)_n} = t_n \end{aligned}$$

and $(A)_n$ is defined in Assumption 2.4.1.

We will first show that $t_n \to t-$ so that X is the pointwise limit of the sequence (X_n), and then that each X_n is $\mathcal{P}$-measurable. This will complete the proof.

Fix t and consider the sequence (t_n). Define $A_t = \cap A_t^n$ ($A_t \in \mathcal{A}$, since $\mathcal{A}$ is assumed to be closed under countable intersections). Note that $A_t^n = (A_t)^n$. This follows since $t \in A_t \Rightarrow A_t^n \subseteq (A_t)^n$. Conversely, $A_t \subseteq A_t^n \in \mathcal{A}_n$ so $(A_t)^n \subseteq A_t^n$. If $t \in A \setminus \cup_{i=1}^k B_i$, $A, B_i \in \mathcal{A}, i = 1, \ldots, k$ (without loss of generality, assume $B_i \subseteq A, \forall i$), then we must show $t_n \in A \setminus \cup_{i=1}^k B_i, \forall n \geq n_0$, some n_0. Now, $A_t^n \subseteq (A \cap A_t)^n$ (since $t \in A$) and so $A_t^n = (A \cap A_t)^n$. This implies that $(A_t^n)_n = ((A \cap A_t)^n)_n \subseteq (A \cap A_t)$ by Assumption

2.4.1. But $t_n \in (A_t^n)_n$, so $t_n \in A$, $\forall n$. It remains to show that for $\forall n$ sufficiently large, $t_n \notin B_i, i = 1, \ldots, k$. Since $t \in A_t, t \notin B_i, i = 1, \ldots, k$, then $A_t \not\subseteq B_i, i = 1, \ldots, k$ and so, by Assumption 2.4.1, there exists n_0 such that $(A_t)_n \not\subseteq B_i, i = 1, \ldots, k, n \geq n_0$. It suffices to show that $(A_t)_n = (A_t^n)_n$. But $(A_t)_n = ((A_t)^n)_n = (A_t^n)_n$ so $t_n \notin B_i, i = 1, \ldots, k, n \geq n_0$, by Assumption 2.4.2. This completes the proof that $t_n \to t-$.

Finally, we must show that X_n is $\mathcal{P}$-measurable, i.e., that $X(\cdot, t_{(A)_n}) I_{G^n(A)}(\cdot)$ is $\mathcal{P}$-measurable. If $G^n(A) = A$, then there does not exist $B \in \mathcal{A}_n, B \neq \emptyset$ such that $B \subset A$. Thus $(A)_n = \emptyset$ and $X(\cdot, t_\emptyset)$ is $\mathcal{F}_\emptyset$-measurable, since X is adapted. If $G^n(A) \neq A$, then we must show that $X(\cdot, t_{(A)_n})$ is measurable with respect to $\mathcal{G}_{G^n(A)} = \cap_{B \in \mathcal{A}, B \cap G^n(A) \neq \emptyset} \mathcal{F}_B = \cap_{B \in \mathcal{A}, B \subseteq A, B \cap G^n(A) \neq \emptyset} \mathcal{F}_B$. Because X is adapted, it suffices to show that $t_{(A)_n} \in B$ if $B \cap G^n(A) \neq \emptyset$, $B \subseteq A$. But $G^n(A) = A \setminus \cup_{A' \subset A, A' \in \mathcal{A}_n} A'$, so we have $B \subseteq A$, $B \not\subseteq A'$, $\forall A' \in \mathcal{A}_n$. Thus we have that $(B)^n = A$ and so $((B)^n)_n = (B)_n = (A)_n$. But $(B)_n \subseteq B$, so $t_{(A)_n} \in B$, as required.

This completes the proof of the theorem. □

Example 2.4.7 This is a very simple example which nevertheless illustrates the fact that T need not be a Hausdorff topological space in order for Theorem 2.4.6 to hold; thus the theorem is a generalization of results of Métivier and Pellaumail (1977) and Allain (1979).

Let $T = [0,1]^2$ and $\mathcal{A} = \{A_x = [0,x] \times [0,1] : 0 \leq x \leq 1\}$. Let $\mathcal{F} = (\mathcal{F}_x) = (\mathcal{F}_{A_x} : A_x \in \mathcal{A})$ be the corresponding filtration. A process $X : \Omega \times [0,1]^2 \to \mathbf{R}$ is adapted if $X(\cdot, (x,y)) \in \mathcal{F}_x$. Let $\mathcal{A}_n = \{[0, k2^{-n}] \times [0,1], k = 0, \ldots, 2^n\}$. This is really a one-dimensional example in a two-dimensional setting. Clearly, $(x_n, y_n) \to (x,y)-$ iff x_n converges to x from below in the usual sense on $[0,1]$. Thus left-convergent sequences have many limits. A function $f : [0,1]^2 \to \mathbf{R}$ is 'left-continuous' if and only if $f(x,y) = f'(x)$, where $f' : [0,1] \to \mathbf{R}$ is left-continuous on $\mathbf{R}$ in the usual sense. The predictable σ-algebra is generated by sets of the form: $\{F \times ((x,1] \times [0,1]), x \in [0,1], F \in \mathcal{F}_x\}$, and also by processes $X : \Omega \times [0,1]^2 \to \mathbf{R}$ which are 'left-continuous' and adapted in the sense defined above.

Comment. As mentioned in Chapter 1, to show that Norberg's (1989b) setting is included in ours, let T be a lattice and let $A_t = \{s \in T, s \leq t\}$. Then there is a one-to-one correspondence $t \leftrightarrow A_t$ and so we may identify $\mathcal{A} = \{A_t : t \in T\}$ with T, and

$A_s \wedge A_t = A_{s \wedge t}, A_s \vee A_t = A_{s \vee t}$. Clearly, t is the unique 'tip' of A_t, so Assumption 2.4.2 is automatically satisfied by $\mathcal{A}$. The classic case in which $T = \mathbf{R}_+^d$ is an example of this simpler structure.

CHAPTER 3

Martingales

In this chapter, we introduce the central notion of martingale. In fact, in the theory of set-indexed martingales, there are three different types of martingales: one is defined using the natural partial order between sets of $\mathcal{A}$, and the two others are defined using the null-increments property of the conditional expectation with respect to the two kinds of past considered.

The theory of the first type of martingale uses essentially the partial order defined on the index set; this theory has already been well developed. The reader can refer to several fundamental papers and books on this subject. Let us mention, in chronological order, Krickeberg (1956), Helms (1958), Chow (1960), Gabriel (1977), Astbury (1978), Millet and Sucheston (1980b), Kurtz (1980a), Washburn and Willsky (1981), Krengel and Sucheston (1981), Hajek and Wong (1981), Hürzeler (1984), and Norberg (1988). A survey of the theory can be found in Edgar and Sucheston (1992).

Almost all these articles are concerned with the problem of convergence for the first type of martingale. If the index set is linearly ordered, under simple assumptions any martingale converges (both almost surely and in L^p-norm). However, the situation becomes much more delicate when the index set is only partially ordered, and we will present some results in this direction. As many fundamental results cannot be generalized to this class of martingales, we must introduce the other two types of martingales mentioned above.

The first steps in the study of the other types of martingales appear in the development of the theory of two-parameter processes with the seminal works of Wong and Zakai (1974) and of Cairoli and Walsh (1975). In order to obtain a satisfactory theory of stochastic integration, these authors introduced strong martingales and weak martingales, as well as other kinds of martingales. These definitions led to a deeper understanding of the concept of martingale, and constituted a starting point for a vast amount of research in the theory of two-parameter processes.

Here, we extend the definitions of strong and weak martingales to the set-indexed framework following Dozzi, Ivanoff and Merzbach (1994) and other papers written later. In the first section, the definitions are given and studied. General classical properties are cursorily mentioned in the second section. Stopping properties are introduced in the third section. A special chapter will be devoted to localization, but the key which permits the study of local martingales is an optional sampling theorem, also called a stopping theorem, which is presented in this section. Finally, applications and examples are given.

3.1 Definitions

Note that we suppose that all the processes $X = \{X_A, A \in \mathcal{A}\}$ defined here are additive (almost surely, cf. Definition 1.4.5). This is satisfied in all the situations we have in mind.

Definition 3.1.1 *Let* $X = \{X_A, A \in \mathcal{A}\}$ *be an adapted and integrable additive process.*

(i) X *is called a* strong martingale *if for all* $C \in \mathcal{C}$, *we have* $E(X_C \mid \mathcal{G}_C^*) = 0$.

(ii) X *is called a* martingale *if for all* $A, B \in \mathcal{A}$ *such that* $A \subseteq B$, *we have* $E(X_B \mid \mathcal{F}_A) = X_A$.

(iii) X *is called a* weak martingale *if for all* $C \in \mathcal{C}$, *we have* $E(X_C \mid \mathcal{G}_C) = 0$.

As usual, if the equality in the definition is replaced by the relation $\geq$ ($\leq$), we add the prefix sub- (super-) before the word 'martingale'. By additivity, without loss of generality it suffices to consider only (sub-) (super-) martingales X with $X_{\emptyset'} = 0$.

Most of the results in this section are from Dozzi, Ivanoff and Merzbach (1994).

Lemma 3.1.2 *Let* $C = A \setminus \cup_{i=1}^n A_i$, $A, A_i \in \mathcal{A}$ *be an extremal representation of* C. *If SHAPE holds, then* $\mathcal{G}_C \subseteq \cap_{i=1}^n \mathcal{F}_{A_i} \cap \mathcal{F}_A$.

Proof. By Lemma 1.1.6, $g_k(A_i) \cap C \neq \emptyset$ whenever the representation of C is extremal. Thus,

$$\mathcal{G}_C \subset \mathcal{F}_{g_k(A_i)} \text{ for all } k \text{ and } i.$$

Therefore, by the outer-continuity of the filtration, $\mathcal{G}_C \subseteq \mathcal{F}_{A_i}$, and since trivially $\mathcal{G}_C \subseteq \mathcal{F}_A$, the result holds. □

DEFINITIONS 55

Lemma 3.1.3 *Let X be a weak submartingale. Then for any C_1 and C_2 in $\mathcal{C}$ such that $C_1 \subseteq C_2$, we have: $0 \leq E(X_{C_1}) \leq E(X_{C_2})$.*

Proof. Note that $C_2 \backslash C_1$ can be written as a finite disjoint union of sets in $\mathcal{C}$:
$$C_2 \backslash C_1 = \cup_{j=1}^k D_j.$$
Therefore:
$$\begin{aligned} E(X_{C_2} - X_{C_1}) &= E(X_{C_2 \backslash C_1}) = E(X_{D_1} + \ldots + X_{D_k}) \\ &= E[E(X_{D_1}|\mathcal{G}_{D_1}) + \ldots + E(X_{D_k}|\mathcal{G}_{D_k})] \geq 0. \end{aligned}$$
□

Notice that a submartingale may be not a weak submartingale. For example, let $X_{[(0,0),z]} = I_{\{(s,t):s+t\geq 1\}}(z)$. Then this process is trivially a submartingale with respect to the deterministic filtration. However $X_{[(\frac{1}{4},\frac{1}{4}),(1,1)]} = -1$. Therefore X is not a weak submartingale.

However, we have the following:

Proposition 3.1.4 *Any strong (sub)martingale is both a (sub)-martingale and a weak (sub)martingale. If SHAPE and (CI) hold, any martingale is a weak martingale.*

Proof. As the first assertion is trivial, we show only the second assertion. Let $C = A \backslash \cup_{i=1}^n A_i$ be an extremal representation of C. Since $\mathcal{G}_C \subseteq \cap_{i=1}^n \mathcal{F}_{A_i} \cap \mathcal{F}_A = \mathcal{F}_{\cap_{i=1}^n A_i \cap A}$ by Lemma 3.1.2 and (CI), we obtain:
$$\begin{aligned} E(X_C|\mathcal{G}_C) &= E[E(X_A - \sum_{i=1}^n X_{A \cap A_i} + \ldots \\ &+ (-1)^n X_{A \cap \cap_{i=1}^n A_i} \mid \mathcal{F}_{\cap_{i=1}^n A_i \cap A})|\mathcal{G}_C] = 0. \end{aligned}$$
□

Now, let us state some properties involving the condition of commutation on the filtration. We begin with the following simple result.

Lemma 3.1.5 *Assume (CI).*

(i) *Let $X = \{X_A, A \in \mathcal{A}\}$ be an integrable and adapted process. Then X is a martingale if and only if for all $A, B \in \mathcal{A}$, $E(X_A|\mathcal{F}_B) = X_{A \cap B}$.*

(ii) *Let X be a martingale and let $A_1, \ldots, A_n \in \mathcal{A}$ be fixed. Then the process Y defined by $Y_A = X_{A \backslash \cup_{i=1}^n A_i}$ is also a martingale.*

Proof. The first part follows from (CI) since by the adaptedness of the process $E(X_A \mid \mathcal{F}_B) = E(X_A \mid \mathcal{F}_{A \cap B})$. Part (ii) follows from part (i). □

Definition 3.1.6 *Let $X = \{X_A, A \in \mathcal{A}\}$ be an integrable and adapted process.*

(i) *X is an L^p-bounded process if $\sup_{C \in \mathcal{C}} E(|X_C|^p) < \infty$, $p \geq 1$.*

(ii) *X is an $\mathcal{A} - L^p$-bounded process if*
$$\|X\|_p = \sup_{A \in \mathcal{A}} \left(E(|X_A|^p)\right)^{1/p} < \infty, \; p \geq 1.$$

Comment: It is important to note that the fact that X is $\mathcal{A} - L^p$-bounded does not necessarily imply that X is L^p-bounded.

It is convenient to write XY for the process defined by the product: $(XY)_A = \{X_A Y_A, A \in \mathcal{A}\}$. The process XX is of course written as X^2. However, for a set $C \in \mathcal{C}$, $(XY)_C$ does not mean $X_C Y_C$, but is defined as in Section 1.4.

For a set $B \in \mathcal{A}$, denote $\mathcal{C}_B = \{A \setminus \cup_{i=1}^n A_i, A, A_i \in \mathcal{A} : B \subseteq A \cap \cap_{i=1}^n A_i\}$.

The next result is similar to a result of Norberg (1989b).

Proposition 3.1.7 *Assume (CI). Let X and Y be two L^2 martingales, $B \in \mathcal{A}$, and $C, D \in \mathcal{C}_B$. Then:*
$$E(X_C Y_D \mid \mathcal{F}_B) = E((XY)_{C \cap D} \mid \mathcal{F}_B).$$

Proof. Let $A, A', A_1, \ldots, A_n$ be elements of $\mathcal{A}$, each containing B. Using Lemma 3.1.5 (i), we have:
$$E(X_A \cdot Y_{A'} \mid \mathcal{F}_B) = E((XY)_{A \cap A'} \mid \mathcal{F}_B).$$

Therefore, by linearity:
$$E(X_{A \setminus \cup_{i=1}^n A_i} \cdot Y_{A'} \mid \mathcal{F}_B) = E((XY)_{(A \cap A') \setminus \cup_{i=1}^n A_i} \mid \mathcal{F}_B).$$

By symmetry, we obtain:
$$E(X_{A \setminus \cup_{i=1}^n A_i} \cdot Y_{A'} \mid \mathcal{F}_B) = E(X_A \cdot Y_{A' \setminus \cup_{i=1}^n A_i} \mid \mathcal{F}_B).$$

Now, using Lemma 3.1.5 (ii), we get the result. □

Proposition 3.1.8 *Assume (CI) and SHAPE. If X is an L^2-martingale, then X^2 is a weak submartingale.*

Proof. Let $C = A \setminus \cup_{i=1}^n A_i$ be maximal in A. Following Proposition 3.1.7, we have: $E((X_C)^2 \mid \mathcal{F}_B) = E(X_C^2 \mid \mathcal{F}_B)$ for any $B \in \mathcal{A}$ such

CLASSICAL PROPERTIES

that $B \subseteq \cap_{i=1}^n A_i$. Choose $B = \cap_{i=1}^n A_i$. Applying Lemma 3.1.2 and (CI), we get:

$$E(X_C^2|\mathcal{G}_C) = E[E(X_C^2|\mathcal{F}_{\cap_{i=1}^n A_i})|\mathcal{G}_C] \geq 0.$$

□

We deal next with monotone outer-continuity. We begin with a very useful result.

Proposition 3.1.9 *Let X and Y be two monotone outer-continuous processes such that for each $A \in \mathcal{A}$, $X_A = Y_A$ a.s. Then X and Y are indistinguishable. That means*

$$X_A = Y_A \quad \text{for all } A \in \mathcal{A} \text{ a.s.}$$
$$\text{Hence} \quad X_C = Y_C \quad \text{for all } C \in \mathcal{C} \text{ a.s.}$$

Proof. The proof is essentially the same as in the classical theory. Since the class $\mathcal{A}' = \cup_n \mathcal{A}_n$ is countable, there exists an event Ω_0 such that $P(\Omega_0) = 1$ and $\forall A \in \mathcal{A}'$ and $\omega \in \Omega_0$ $X_A(\omega) = Y_A(\omega)$. We then can conclude by monotone outer-continuity in $\mathcal{A}$. □

Proposition 3.1.10 *For $p \geq 1$, every L^p-martingale is L^p-monotone outer-continuous on $\mathcal{A}$.*

Proof. Let X be a L^p-martingale and $\{A_n\}$ a decreasing sequence from the class $\mathcal{A}$ converging to A. For $m \leq n$, we have:

$$E(X_{A_m}|\mathcal{F}_{A_n}) = X_{A_n}.$$

Thus, the sequence $(Y_n, \mathcal{F}_n) = (X_{A_n}, \mathcal{F}_{A_n})$ is a reverse martingale, with

$$E(Y_n) = E(X_A) > -\infty.$$

Applying the reverse martingale convergence theorem (cf. Kopp (1984), Corollary 2.10.2), it follows that

$$X_{A_n} = Y_n \to Y_\infty = X_A \text{ in } L^p\text{-norm}.$$

□

3.2 Classical properties

In this section, we mention very briefly some classical properties of martingales which can be easily extended to set-indexed martingales. We begin with decompositions; however, due to their importance, a special chapter will be devoted to Doob-Meyer decompositions. Then we discuss different kinds of convergence for set-indexed martingales, which are consequences of the theory of martingales indexed by directed sets.

Let (A_n) be an increasing sequence in $\mathcal{A}$. We write $A_n \uparrow T$ if $A_n \neq T$, $\forall n$ and if $\overline{\cup A_n} = T$.

Definition 3.2.1 *A positive supermartingale $\{X_A, A \in \mathcal{A}\}$ is called a* potential *if $\lim_{A\uparrow T} EX_A = 0$. (For any increasing sequence $\{A_n\}$ in $\mathcal{A}$ such that $\overline{\cup A_n} = T$, we have $\lim_{n\to\infty} EX_{A_n} = 0$.)*

For the remainder of this section the following must be assumed:

Assumption 3.2.2 *$T \notin \mathcal{A}$, but there exists a sequence (A_n) in $\mathcal{A}$ such that $A_n \uparrow T$. Also, if (B_n) is any sequence in $\mathcal{A}$ such that $B_n \uparrow T$, then for any $A \in \mathcal{A}$ there exists n_A such that $A \subseteq B_n$, $\forall n \geq n_A$.*

The following results from Burstein (1999) are extensions of classical results (see for example Dellacherie and Meyer (1980)).

Theorem 3.2.3 (Riesz decomposition)
Assume that SHAPE holds. Every positive supermartingale $\{X_A, A \in \mathcal{A}\}$ (or more generally, any supermartingale such that $\lim_{A\uparrow T} E|X_A| > -\infty$) has a decomposition of the form $X = Y + Z$, where Y is a martingale and Z is a potential. This decomposition is unique and Y is the greatest submartingale bounded above by X.

Proof. For every $A, B, C \in \mathcal{A}, A \subseteq B \subseteq C$ we have:

$$X_A \geq E(X_B|\mathcal{F}_A) \geq E(E(X_C|\mathcal{F}_B)|\mathcal{F}_A) = E(X_C|\mathcal{F}_A),$$

i.e. the sequence $\{E(X_B|\mathcal{F}_A)\}$ decreases as $B \uparrow T$. Let Y_A be its limit which is unique a.s. because of Assumption 3.2.2. As $E(X_B|\mathcal{F}_A)$ is bounded above by X_A which is integrable, Y_A is integrable if and only if $\lim_{B\uparrow T} E(E(X_B|\mathcal{F}_A)) > -\infty$. This condition is independent of A and means simply that $\lim_{B\uparrow T} E(X_B) > -\infty$. So, Y_A is integrable. Then

$$Y_A = \lim_{B\uparrow T} E(X_B|\mathcal{F}_A) = \lim_{B\uparrow T} E(E(X_B|\mathcal{F}_{A'})|\mathcal{F}_A) = E(Y_{A'}|\mathcal{F}_A),$$

where $A \subseteq A' \subseteq B$. So, the process Y_A is a martingale (additivity follows from SHAPE). Define $Z_A = X_A - Y_A$. We have seen already that $Z_A \geq 0$; it is a supermartingale as a sum of two supermartingales X_A and $-Y_A$. On the other hand,

$$EZ_A = EX_A - EY_A = EX_A - E\lim_{B\uparrow T} E(X_B|\mathcal{F}_A) = EX_A - \lim_{B\uparrow T} EX_B,$$

so $EZ_A \to 0$ as $A \uparrow T$. Consequently, Z is a potential. Now, if H is a submartingale bounded above by X, then, since $H_B \leq X_B$ $\forall B \in$

CLASSICAL PROPERTIES 59

$\mathcal{A}$, $H_A \le E(H_B|\mathcal{F}_A) \le E(X_B|\mathcal{F}_A) \to Y_A$ for $A \subseteq B$, so $H \le Y$ and Y is the greatest submartingale bounding X from below.

To show the uniqueness of a decomposition, assume that $X_A = Y_A + Z_A = Y'_A + Z'_A$, then $Y_A - Y'_A = Z'_A - Z_A$. This is a martingale as a difference of two martingales. It is sufficient to show that a martingale M which is also the difference of two potentials is zero. But the process $|M|$ is a submartingale, so $|M_A| \le E(|M_B| \mid \mathcal{F}_A), A \subseteq B$ and $E|M_A| \le \lim_{B \uparrow T} E(|M_B|) = 0$, which implies $M_A = 0$ a.s. for all $A \in \mathcal{A}$. So, $Y_A = Y'_A$ and $Z_A = Z'_A$. □

Theorem 3.2.4 (Krickeberg decomposition)
Assume that SHAPE holds. A martingale $\{X_A, A \in \mathcal{A}\}$ is $\mathcal{A}$-bounded in L^1 if and only if it is the difference of two positive martingales, and then it has a unique decomposition of the form $X = Y - Z$, where Y and Z are two positive martingales such that $||X||_1 = ||Y||_1 + ||Z||_1$, where $||X||_1 = \sup_{A \in \mathcal{A}} E|X_A|$. Moreover, Y is the smallest positive supermartingale bounding X above and Z the smallest positive supermartingale bounding $-X$ above.

Proof. Sufficiency is trivial: every positive martingale is $\mathcal{A}$-bounded in L^1 and so is every difference of two positive martingales. Conversely, let X be a martingale which is $\mathcal{A}$-bounded in L^1. Define the processes $X^+ = \max(X, 0)$, $X^- = -\min(X, 0)$. The process $-X_A^+$ is a supermartingale, such that $\lim_{A \uparrow T} E(-X_A^+) > -\infty$. Hence it has the Riesz decomposition $-X_A^+ = -Y_A + P_A$, where $-Y$ is a martingale, P is a potential, and we have seen that $-Y_A = \lim_{B \uparrow T} E(-X_B^+ \mid \mathcal{F}_A)$. Rewrite it as $X_A^+ = Y_A - P_A$. Similarly, for a supermartingale $-X_A^-$ we can write $-X_A^- = -Z_A + Q_A$, where Q is a potential and the martingale $-Z$ is given by $-Z_A = \lim_{B \uparrow T} E(-X_B^- \mid \mathcal{F}_A)$, and rewrite it as $X_A^- = Z_A - Q_A$. Clearly, Y and Z are positive martingales, and we have

$$Y_A - Z_A = \lim_{B \uparrow T} E(X_B^+ - X_B^- \mid \mathcal{F}_A) = \lim_{B \uparrow T} E(X_B \mid \mathcal{F}_A) = X_A,$$

so that $X = Y - Z$, the difference of two positive martingales.

We have

$$\begin{aligned}
||Y||_1 + ||Z||_1 &= \lim_{A \uparrow T} E|Y_A| + \lim_{A \uparrow T} E|Z_A| \\
&= \lim_{A \uparrow T} EY_A + \lim_{A \uparrow T} EZ_A \\
&= E(Y_{\emptyset'} + Z_{\emptyset'}) = \lim_{B \uparrow T} EX_B^+ + \lim_{B \uparrow T} EX_B^- \\
&= \lim_{B \uparrow T} E(X_B^+ + X_B^-)
\end{aligned}$$

$$= \lim_{B \uparrow T} E|X_B| = ||X||_1.$$

It follows from the Riesz decomposition that $-Y$ is the greatest submartingale bounded above by $-X^+$, so Y is the smallest positive supermartingale bounding X^+ (and therefore X) above. The analogous assertion for Z follows by replacing X by $-X$.

Now consider uniqueness: if we write X as the difference of two positive martingales $X = U - V$, then $U \geq X^+, V \geq X^-$. By the Riesz decomposition, $U \geq Y, V \geq Z$. If $||X||_1 = ||U||_1 + ||V||_1 = ||Y||_1 + ||Z||_1$, then $||U||_1 = ||Y||_1, ||V||_1 = ||Z||_1$. On the other hand, U, Y, and $U - Y$ are positive martingales, therefore $||U||_1 = ||Y||_1 + ||U - Y||_1$; hence $||U - Y||_1 = 0$, and finally $U = Y$ and $V = Z$. □

Now we briefly discuss convergence properties.

As noted in Walsh (1986b), the convergence theorems for martingales having a general partially ordered index set offer a striking contrast. Those involving convergence in probability and in L^p are easy consequences of the usual (real-indexed) martingale convergence theorems, while the extensions of the usual almost everywhere convergence theorems break down entirely (since there is no metric corresponding to a.e. convergence), and can only be salvaged by putting new conditions on either the σ-fields or on the parameter set.

Helms (1958) gave necessary and sufficient conditions for mean convergence in L^p for martingales indexed by a directed set. Vitali conditions of different types are an example of such restrictions on σ-algebras. They are introduced to ensure essential convergence of martingales. The first paper which deals with them was written by Krickeberg (1956), who used a Vitali condition and proved the a.e. convergence of separable martingales indexed by a directed set, with bounded variation.

This condition was generalized by Chow (1960) and later by Millet and Sucheston (1980b) and others for the two-parameter case. We refer to Burstein (1999) for more details. Summing up this topic, we have the following classic result from Neveu (1975).

Theorem 3.2.5 *Let $\{X_A, A \in \mathcal{A}\}$ be a martingale in L^p ($1 \leq p < \infty$). Then the following are equivalent:*

(i) $\lim X_A$ *exists along $\mathcal{A}$ in L^p. (There exists a random variable such that for any L^p-neighbourhood S of it, there is a set $A \in \mathcal{A}$ such that $A \subseteq B, B \in \mathcal{A}$ implies $X_B \in S$.)*

(ii) *The family $\{X_A, A \in \mathcal{A}\}$ is relatively compact, weakly in L^p (i.e., if $p = 1$: uniformly integrable; if $p > 1$: bounded).*

(iii) *There exists a random variable X in L^p such that $\forall A : X_A = E(X \mid \mathcal{F}_A)$.*

These conditions imply that:

$$X_A = E(\lim_{B \uparrow T} X_B \mid \mathcal{F}_A) \text{ a.s.}$$

3.3 Stopping theorems

A first stopping theorem for martingales indexed by points of a directed set was obtained by Kurtz (1980a) and extended to a very special class of submartingales by Hürzeler (1985b) and others. A related topic is the theory of optimal stopping, but we will not deal with it here (see for example Nualart (1992)). The results presented here come essentially from Ivanoff and Merzbach (1995); they extend previous results obtained in the eighties for the two-parameter case.

This section is divided as follows: we deal first with strong (sub-)martingales, next the (sub-)martingale case is discussed and finally some stopping theorems are given for weak (sub-)martingales.

In this section, since approximations of stopping sets and σ-algebras generated by stopping sets are used (as defined in Section 1.5), we assume that SHAPE or the condition of Lemma 1.3.4 is satisfied.

We begin with an optional sampling theorem which characterizes the strong martingale property.

Theorem 3.3.1 *Assume that $\mathcal{F}_B = \mathcal{F}_B^r \ \forall B \in \mathcal{A}(u)$. Let X be adapted and monotone outer-continuous. Then X is a strong (sub-)martingale if and only if for any two stopping sets ξ, ξ' we have*

$$E(X_{\xi'} \mid \mathcal{F}_\xi) = X_{\xi' \cap \xi} \ (\geq X_{\xi' \cap \xi}) \text{ a.s.} \tag{3.1}$$

Proof. We prove the submartingale case only, as the proof for strong martingales is identical, replacing all inequalities in what follows with equalities. First we assume that the inequality defined in (3.1) is satisfied and let $C = A \setminus D \in \mathcal{C}$, $A \in \mathcal{A}$, $D \in \mathcal{A}(u)$. To show that X is a strong submartingale, since it is easily seen that $\cup_{B \in \mathcal{A}(u), B \cap C = \emptyset} \mathcal{F}_B$ is a π-system generating $\mathcal{G}_C^*$, it suffices to show that $\int_F X_C dP \geq 0$ for $F \in \mathcal{F}_B$, some $B \in \mathcal{A}(u)$, $B \cap C = \emptyset$. Since

$B \cap C = \emptyset$, $C = A \setminus (D \cup B)$. Let $\xi' = A$ and $\xi = D \cup B$ in (3.1). Since $F \in \mathcal{F}_B \subseteq \mathcal{F}_{(D \cup B)}$,

$$\int_F X_C dP = \int_F X_A dP - \int_F X_{(D \cup B) \cap A} dP$$
$$= \int_F E(X_A \mid \mathcal{F}_{(D \cup B)}) dP - \int_F X_{(D \cup B) \cap A} dP$$
$$\geq 0.$$

To prove the converse, assume that X is a strong submartingale. Recall that $\xi \cap \xi'$ is a stopping set.

We now begin by assuming that ξ, ξ' are both discrete and bounded by $A \in \mathcal{A}(u)$. As proven in Lemma 2.1.5, we may assume that there exists a finite subsemilattice $\mathcal{A}'$ of $\mathcal{A}$, such that ξ and ξ' take their values in $\mathcal{A}'(u)$, and if $C^\ell(\mathcal{A}')$ is the class of left-neighbourhoods of $\mathcal{A}'$ (say $C^\ell(\mathcal{A}') = \{C_1, \ldots, C_n\}$), then:

$$(\emptyset, \xi'] = \bigcup_{i=1}^n F'_i \times C_i \text{ and } (\emptyset, \xi' \cap \xi] = \bigcup_{i=1}^n F_i \times C_i,$$

where $F'_i = \{\omega : C_i \subseteq \xi'(\omega)\}$, $F_i = \{\omega : C_i \subseteq (\xi \cap \xi')(\omega)\}$, $F_i \subseteq F'_i$, $i = 1, \ldots, n$, and $F_i, F'_i \in \mathcal{G}_{C_i}$.

Let $K \in \mathcal{F}_\xi$. Then $K \cap \{\xi = B\} \in \mathcal{F}_B^r$. We have

$$(K \times A) \cap (\xi \cap \xi', \xi'] = \bigcup_{\substack{B \in \mathcal{A}'(u) \\ }} \bigcup_{\substack{C_i \not\subseteq B \\ C_i \in C^\ell(\mathcal{A}')}} (K \cap \{\xi = B\} \cap F'_i) \times C_i.$$

But since $C_i \in C^\ell(\mathcal{A}')$, $C_i \not\subseteq B$ implies that $B \cap C_i = \emptyset$. Thus, $\mathcal{F}_B \in \mathcal{G}^*_{C_i}$, and since $F'_i \in \mathcal{G}_{C_i} \subseteq \mathcal{G}^*_{C_i}$, we have that $K \cap \{\xi = B\} \cap F'_i \in \mathcal{G}^*_{C_i}$. Therefore,

$$E[(X_{\xi'} - X_{\xi \cap \xi'}) I_K]$$
$$= E\left[\bigcup_{\substack{B \in \mathcal{A}'(u)}} \bigcup_{\substack{C_i \not\subseteq B \\ C_i \in C^\ell(\mathcal{A}')}} X_{C_i} I_{\{\xi = B\} \cap F'_i} I_K \right]$$
$$= E\left[\bigcup_{\substack{B \in \mathcal{A}'(u)}} \bigcup_{\substack{C_i \not\subseteq B \\ C_i \in C^\ell(\mathcal{A}')}} I_{K \cap \{\xi = B\} \cap F'_i} E(X_{C_i} \mid \mathcal{G}^*_{C_i}) \right]$$
$$\geq 0.$$

Thus, $E(X_{\xi'} \mid \mathcal{F}_\xi) \geq X_{\xi \cap \xi'}$.

Now assume that ξ' and ξ are not necessarily discrete, and that $\xi \subseteq \xi' \subseteq A$, $A \in \mathcal{A}(u)$. Consider $E(X_{g_n(\xi')} \mid \mathcal{F}_{g_m(\xi)})$, for $m \geq n$.

STOPPING THEOREMS 63

For n fixed, $(E(X_{g_n(\xi')}|\mathcal{F}_{g_m(\xi)}), \mathcal{F}_{g_m(\xi)})$ is a reverse martingale. Therefore (cf. Kopp (1984), Corollary 2.10.2), as $m \to \infty$

$$E(X_{g_n(\xi')}|\mathcal{F}_{g_m(\xi)}) \to E(X_{g_n(\xi')}|\mathcal{F}_\xi)$$

a.s. and in L^1 (since $\cap_m \mathcal{F}_{g_m(\xi)} = \mathcal{F}_\xi$). Next, it is easily seen by what was proven above that $(X_{g_n(\xi')}, \mathcal{F}_{g_n(\xi')})$ is a reverse submartingale, and $0 \leq E(X_{g_n(\xi')}) \leq E(X_{g_1(A)})$ $\forall n$. Thus $(X_{g_n(\xi')})$ is uniformly integrable (Kopp (1984), Theorem 2.10.1), and as $n \to \infty$, $X_{g_n(\xi')} \to X_{\xi'}$ a.s. and in L^1. Thus we also have $E(X_{g_n(\xi')}|\mathcal{F}_\xi) \to E(X_{\xi'}|\mathcal{F}_\xi)$ a.s. and in L^1. Finally, it follows that

$$\begin{aligned} 0 &\leq \lim_{n\to\infty} \lim_{m\to\infty} (E(X_{g_n(\xi')}|\mathcal{F}_{g_m(\xi)}) - X_{g_m(\xi\cap\xi')}) \\ &= E(X_{\xi'}|\mathcal{F}_\xi) - X_{\xi\cap\xi'}, \text{a.s.} \end{aligned}$$

This completes the proof. $\square$

We say that any adapted process X is of class D if the family $\{X_\xi : \xi \in \tau^f\}$ is uniformly integrable, where τ^f is the class of discrete stopping sets. The proof of the following corollary is analogous to the classic case.

Corollary 3.3.2 *If $T \in \mathcal{A}$, then every positive (strong) submartingale is of class D.*

We now consider the martingale case and begin, for the sake of completeness, with a result due to T.G. Kurtz (1980a):

Theorem 3.3.3 *Let X be an adapted and integrable process and let α and α' be two simple stopping sets such that $\alpha \subseteq \alpha'$. Then X is a martingale if and only if*

$$E(X_{\alpha'} \mid \mathcal{F}_\alpha) = X_\alpha. \tag{3.2}$$

Comment. It is clear from Theorem 3.3.1 that this result cannot be extended to general stopping sets (and therefore, in general, a martingale is not of class D). However, we have the following:

Theorem 3.3.4 *Assume SHAPE and (CI). Let X be a monotone outer-continuous martingale and ξ be a stopping set. If X is of class D or if ξ is discrete then for any $A, B \in \mathcal{A}$,*

$$E(X_{A\cap\xi} \mid \mathcal{F}_B) = X_{B\cap\xi}. \tag{3.3}$$

If (CI) does not hold, then (3.3) is valid for any $A, B \in \mathcal{A}$ such that $B \subseteq A$.

Proof. Assume first (CI) and that ξ is discrete. Using the representation of Lemma 2.1.5, we can write:

$$E(X_{A \cap \xi} \mid \mathcal{F}_B) = \sum_i E(X_{A \cap C_i} I_{F_i} \mid \mathcal{F}_B),$$

where $F_i = \{C_i \subseteq \xi\} \in \mathcal{G}_{c_i}$, and the $\{C_i\}$ are disjoint. Also, without loss of generality, we may assume that $C_i \not\subseteq B$ implies $C_i \cap B = \emptyset$. Since X is a martingale and (CI) is assumed, this formula is equal to:

$$\sum_{\substack{i \\ C_i \subseteq B}} X_{C_i} I_{F_i} + \sum_{\substack{i \\ C_i \not\subseteq B}} E(X_{A \cap C_i} I_{F_i} \mid \mathcal{F}_B).$$

The first summation is exactly $X_{B \cap \xi}$. It remains to show that the second summation vanishes. Note that since $C_i \not\subseteq B$, we have $C_i \cap B = \emptyset$. Let $C_i \cap A = A^{(i)} \setminus \cup_j A_j^{(i)}$ be a maximal representation of $C_i \cap A$ in $g_1(D(A, B))$, where $D(A, B)$ is any set in $\mathcal{A}(u)$ such that $A \subseteq D(A, B)$ and $B \subseteq D(A, B)$. (The existence of such a set is ensured by #1 of Definition 1.1.1.) Therefore, by SHAPE, among all the sets $\{A_j^{(i)}\}_j$, there is at least one, for example $A_1^{(i)}$, such that $B \subseteq A_1^{(i)}$. Now recall that

$$X_{C \cap A} = X_{A^{(i)}} - \sum_i X_{A \cap A_i} + \sum_{i<j} X_{A \cap A_i \cap A_j} + \ldots + (-1)^n X_{A \cap A_1 \cap \ldots \cap A_n}.$$

(The sub- and superscript (i) has been suppressed for notational convenience.) By (CI), $E(X_{A^{(i)}} \mid \mathcal{F}_{A_1}) = E(X_{A^{(i)} \cap A_1})$, $E(X_{A^{(i)} \cap A_i} \mid \mathcal{F}_{A_1}) = E(X_{A^{(i)} \cap A_1 \cap A_i})$, if $i \neq 1$, etc. Thus, the sum telescopes, and $E(X_{C_i \cap A} \mid \mathcal{F}_{A_1^{(i)}}) = 0$. Now, since $g_n(A_j^{(i)}) \cap C_i \neq \emptyset \; \forall n$, we have $\mathcal{G}_{C_i} \subseteq \mathcal{F}_{A_j^{(i)}} \; \forall j$. Thus, if $B \subseteq A_j^{(i)}$,

$$E(X_{A \cap C_i} I_{F_i} \mid \mathcal{F}_B) = E[I_{F_i} E(X_{A \cap C_i} \mid \mathcal{F}_{A_j^{(i)}}) \mid \mathcal{F}_B] = 0.$$

Thus, $E(X_{A \cap \xi} \mid \mathcal{F}_B) = X_{B \cap \xi}$, if ξ is discrete.

If X is of class D, observe that by uniform integrability and L^1-convergence,

$$\begin{aligned} E(X_{A \cap \xi} \mid \mathcal{F}_B) &= \lim_n E(X_{A \cap g_n(\xi)} \mid \mathcal{F}_B) \\ &= \lim_n X_{B \cap g_n(\xi)} \\ &= X_{B \cap \xi}. \end{aligned}$$

STOPPING THEOREMS   65

If (CI) does not hold, and $B \subseteq A$, then the above proof is still valid, replacing $\mathcal{F}_{A_j^{(i)}}$ with $\mathcal{F}_{A \cap A_j^{(i)}}$. □

Finally, we come to the weak martingale case, and state first a general result.

Proposition 3.3.5 *Let X be a L^1-outer-continuous weak (sub-) martingale of class D. Then for any two stopping sets σ and τ such that $\sigma \subseteq \tau$, we have*

$$E(X_\tau - X_\sigma) = 0 \ (\geq 0). \tag{3.4}$$

Proof. If σ and τ are both discrete stopping sets, from Lemma 2.1.5 the remark following, we may assume that there exist finitely many disjoint sets $C_1, \ldots, C_k \in \mathcal{C}$ such that $(\emptyset, \tau] = \cup_{j=1}^k F_j \times C_j$, $(\emptyset, \sigma] = \cup_{j=1}^k G_j \times C_j$, where $F_j, G_j \in \mathcal{G}_{C_j}$ and $G_j \subseteq F_j, j = 1, \ldots, k$. Then

$$\begin{aligned}(\sigma, \tau] &= (\cup_{j=1}^k F_j \times C_j) \setminus (\cup_{j=1}^k G_j \times C_j) \\ &= \cup_{j=1}^k (F_j \times C_j) \setminus (G_j \times C_j) \text{ since the } C_j\text{'s are disjoint} \\ &= \cup_{j=1}^k (F_j \setminus G_j) \times C_j.\end{aligned}$$

Thus,

$$E(X_\tau - X_\sigma) = E[\sum_{j=1}^k I_{F_j \setminus G_j} E(X_{C_j} \mid \mathcal{G}_{C_j})] = 0 \ (\geq 0).$$

Now for general stopping sets, by Lemmas 1.5.4 and 1.5.6, σ and τ may be approximated by above by decreasing sequences of discrete stopping sets. Therefore, by uniform integrability and L^1-monotone outer-continuity of X, we obtain (3.4). □

Corollary 3.3.6 *Under the assumptions of Proposition 3.3.5, we have*

$$E[(X_\tau - X_\sigma) \mid \mathcal{G}(\sigma, \tau)] = 0 \ (\geq 0), \tag{3.5}$$

where $\mathcal{G}(\sigma, \tau) = \sigma\{F : F \in \mathcal{F}, \sigma_F(\tau) \text{ is stopping set}\}$, and

$$\sigma_F(\tau)(\omega) = \begin{cases} \sigma(\omega) & \text{if } \omega \in F \\ \tau(\omega) & \text{if } \omega \notin F \end{cases}$$

Proof. $\mathcal{G}^0(\sigma, \tau) = \{F : F \in \mathcal{F}, \sigma_F(\tau) \text{ is a s.s.}\}$ is a field (and therefore a π-system) and following Proposition 3.3.5, for any $F \in \mathcal{G}^0(\sigma, \tau)$, we obtain $E[(X_\tau - X_\sigma) \cdot I_F] = E[X_\tau - X_{\sigma_F(\tau)}] = 0 \ (\geq 0)$. Therefore (3.5) holds. □

Comment 3.3.7 In general, it is difficult to characterize the σ-field $\mathcal{G}(\sigma,\tau)$ and in certain cases it is the trivial σ-field $\mathcal{F}_\emptyset$. This shows that we cannot expect to get an interesting optional sampling theorem for weak (sub-)martingales. However, notice that if $T \in \mathcal{A}$ and $\tau = T$ is deterministic, we have $\mathcal{G}(\sigma,\tau) = \{F \in \mathcal{F} : F \cap \{A \not\subseteq \sigma\} \in \mathcal{F}_A, \forall A \in \mathcal{A}\}$.

The next result is similar to Theorem 3.3.4.

Theorem 3.3.8 *Let X be a monotone outer-continuous weak (sub-)martingale and ξ be a stopping set. If X is of class D or if ξ is discrete, then*

$$E(X_{C \cap \xi} \mid \mathcal{G}_C) = 0 \quad (\geq 0) \quad \text{for any } C \in \mathcal{G}. \tag{3.6}$$

Proof. Suppose first that ξ is discrete and following Lemma 2.1.5, let $(\emptyset, \xi] = \cup_{i=1}^k F_i \times C_i$. Now, $X_{C \cap \xi} = \sum_i I_{F_i} X_{C \cap C_i}$ and note that $\mathcal{G}_C \subseteq \mathcal{G}_{C \cap C_i}$ and $\mathcal{G}_{C_i} \subseteq \mathcal{G}_{C \cap C_i}$. Then

$$\begin{aligned} E(X_{C \cap \xi} \mid \mathcal{G}_C) &= \sum_i E[E(X_{C \cap C_i} I_{F_i} \mid \mathcal{G}_{C \cap C_i}) \mid \mathcal{G}_C] \\ &= \sum_i E[I_{F_i} E(X_{C \cap C_i} \mid \mathcal{G}_{C \cap C_i}) \mid \mathcal{G}_C] = 0 \quad (\geq 0). \end{aligned}$$

For a general stopping set ξ, since $\xi = \cap_n g_n(\xi)$ which are discrete stopping sets, using the fact that X is a.s. monotone outer-continuous and uniform integrability (class D property), we obtain the result. $\square$

3.4 Examples

In this section we will be considering the set-indexed Brownian motion and Poisson processes. We will observe that both of these processes have *independent increments*: an $\mathcal{A}$-indexed process X has independent increments if $X_{C_1}, ..., X_{C_n}$ are independent random variables whenever $C_1, ..., C_n$ are disjoint sets in $\mathcal{C}$. We begin with the following general result:

Theorem 3.4.1 *Let X be an (additive) $\mathcal{A}$-indexed process with mean 0 independent increments (i.e. $E(X_C) = 0$, $\forall C \in \mathcal{C}$). Then X is a strong martingale with respect to its minimal filtration $\{\mathcal{F}_A^0 : A \in \mathcal{A}\}$ ($\mathcal{F}_A^0 = \sigma\{X_{A'} : A' \in \mathcal{A}, A' \subseteq A\}$). If X is monotone outer-continuous and uniformly integrable on B_h for each h (i.e. $(X_A : A \in \mathcal{A}, A \subseteq B_h)$ is uniformly integrable, B_h as in Definition*

EXAMPLES 67

1.1.1(1))*, then X is a strong martingale with respect to its minimal outer-continuous filtration.*

Proof. The first statement follows trivially from independence of the increments.

Now suppose that $\mathcal{F}_B = \mathcal{F}_B^r$, for $B \in \mathcal{A}(u)$. Let $C \in \mathcal{C}$. Let $C = A \setminus \cup_{i=1}^k A_i^h = A \setminus A^h$ be the maximal representation of C in B_h, $A, A_i^h \in \mathcal{A}$, $A^h \in \mathcal{A}(u)$, $h \in \mathbf{N}$. In what follows, we shall always assume that n, m are sufficiently large that if $A \subset T$, $g_n(A)$ $(g_m(A)) \subset T$.

Let $C_m = A \setminus \cup_{i=1}^k g_m(A_i^h)$ be an approximation of C. $(C_m)_m$ is an increasing sequence which converges to C, so by monotone outer continuity and uniform integrability, $(X_{C_m})_m$, converges to X_C, a.s. and in L^1. Also, for each m,

$$g_n(A^h) \cap C_m = \emptyset \text{ for all } n > m.$$

Then, by independence of the increments of X, for n sufficiently large,

$$E(X_{C_m} \mid \mathcal{F}^0_{g_n(A^h)}) = E(X_{C_m}) = 0.$$

On the other hand, by the reverse martingale convergence theorem,

$$\lim_{n \to \infty} E(X_{C_m} \mid \mathcal{F}^0_{g_n(A^h)}) = E(X_{C_m} \mid \cap_{n=1}^\infty \mathcal{F}^0_{g_n(A^h)}) = 0.$$

Since $(X_{C_m})_m$ converges to X_C in L^1, then

$$\begin{aligned}
E(X_C \mid \mathcal{F}^r_{A^h}) &= E(X_C \mid \cap_{n=1}^\infty \mathcal{F}^0_{g_n(A^h)}) \\
&= \lim_{m \to \infty} E(X_{C_m} \mid \cap_{n=1}^\infty \mathcal{F}^0_{g_n(A^h)}) \\
&= 0.
\end{aligned}$$

It is easily seen that $\mathcal{G}_C^* = \vee_h \mathcal{F}^r_{A^h}$. But the sequence $(E(X_C \mid \mathcal{F}^r_{A^h}))_h$ is a uniformly integrable martingale, so

$$E(X_C \mid \mathcal{G}_C^*) = \lim_{h \to \infty} E(X_C \mid \mathcal{F}^r_{A^h}) = 0, \text{ a.s.}$$

□

We continue with **Brownian motion**.

Definition 3.4.2 *Let Λ be a non-negative increasing function defined on $\mathcal{A}$ with $\Lambda_{\emptyset'} = 0$. We say that an $\mathcal{A}$-indexed process X is a Brownian motion with variance measure Λ if $X_{\emptyset'} = 0$, if X can be extended to an additive process on $\mathcal{C}(u)$, and if for disjoint sets $C_1, \ldots, C_n \in \mathcal{C}, X_{C_1}, \ldots, X_{C_n}$ are independent mean-zero Gaussian random variables with variances $\Lambda_{C_1}, \ldots \Lambda_{C_n}$, respectively.*

Comment. Given the increasing function Λ, the question of the existence of X arises. Since each of the sets in $\mathcal{A}$ is compact, then for each $A \in \mathcal{A}$, Λ defines a unique finite measure on $\sigma(B \in \mathcal{A}, B \subseteq A)$. Thus, an appropriate (almost surely) additive set-indexed Gaussian process exists on $\sigma(B \in \mathcal{A}, B \subseteq A)$, which can be extended to all sets in $\mathcal{A}$ by Kolmogorov's extension theorem (cf. Adler (1990)).

Remarks 3.4.3 1. By Theorem 3.4.1, X is a strong martingale with respect to $\{\mathcal{F}_A^0, A \in \mathcal{A}\}$ and with respect to $\{\mathcal{F}_A^r, A \in \mathcal{A}\}$ if it is monotone outer-continuous. In addition, it will be shown in the next chapter that if X is a monotone inner- and outer-continuous set-indexed Brownian motion with variance measure Λ such that $X_{\emptyset'} = 0$, then for every $C \in \mathcal{C}(u)$, $E[(X_C)^2 \mid \mathcal{G}_C^*] = \Lambda_C$, for $\mathcal{G}_C^*$ defined either using $\mathcal{F}^0$ or $\mathcal{F}^r$. Subsequently, it will be shown that this property characterizes the set-indexed Brownian motion.

2. If X is a monotone inner- and outer-continuous set-indexed Brownian motion with variance measure Λ, then Λ is also monotone inner- and outer-continuous. This follows immediately by uniform integrability. Conversely, conditions under which monotone inner- and outer-continuity of Λ ensure the existence of a monotone inner- and outer-continuous version of X are thoroughly discussed in Adler (1990). In particular, it is easily shown that it suffices that $\mathcal{A}$ be a Vapnik-Červonenkis class in $\sigma(\mathcal{A})$.

However, if $\mathcal{A}$ is too large, the Brownian motion process can be very badly behaved, as shown by the next example due to Dudley (1979). If $T = [0,1]^2$, then $W = \{W_A, A \in \mathcal{A}\}$ is unbounded and discontinuous on the 'lower sets' in T (as defined in Example 1.2.4). The following proof is as presented in Adler (1990):

Writing a point in $[0,1]^2$ as (s,t), let T_{01} be the right triangle in which $s \leq 1$ and $t \leq 1 \leq s+t$. Let C_{01} be the square where $\frac{1}{2} < s \leq 1$ and $\frac{1}{2} \leq t \leq 1$.

For $n = 1, 2, \ldots$, and $j = 1, \ldots, 2^n$, let T_{nj} be the right triangle defined by $s + t \geq 1, (j-1)2^{-n} \leq s < j2^{-n}$, and $1 - j2^{-n} < t \leq 1 - (j-1)2^{-n}$. Let C_{nj} be the square filling the upper right corner of T_{nj}, in which $(2j-1)2^{-(n+1)} \leq s < j2^{-n}$ and $1 - (2j-1)2^{-(n+1)} \leq t < 1 - (j-1)2^{-n}$.

Since the squares C_{nj} are disjoint for all n and j, the random variables $W(C_{nj})$ are independent. We have $\lambda(C_{nj}) = 4^{-(n+1)}$ for all n, j, where λ is the Lebesgue measure.

EXAMPLES

Let D be the negative diagonal $\{(s,t) \in [0,1]^2 : s+t = 1\}$, and $L_{nj} = D \cap T_{nj}$. For each $n \geq 1$, each point $p = (s,t) \in D$ belongs to exactly one such interval $L_{n,j(n,p)}$ for some unique $j(n,p)$.

For each $p \in D$ and $M < +\infty$ the events

$$E_{np} = \{W(C_{n,j(n,p)}) > M 2^{-(n+1)}\}$$

are independent for $n = 0, 1, 2, \ldots$, and have the same positive probability (since $W(C_{nj})/2^{-(n+1)}$ is standard normal for all n and j). Thus, by the Borel-Cantelli lemma, such an event occurs with probability one. Let $n(p) = n(p, \omega)$ be the least such n, defined and finite for almost all ω.

Since the events $E_{np}(\omega)$ are measurable jointly in p and ω, Fubini's theorem implies that, with probability one, for almost all p (with respect to Lebesgue measure on D) some E_{np} occurs, and $n(p) < +\infty$. Let

$$V_\omega = \cup_{p \in D} T_{n(p), j(n(p), p)},$$
$$A_\omega = \{(s,t) : s+t \leq 1\} \cup V_\omega,$$
$$B_\omega = A_\omega \setminus \cup_{p \in D} C_{n(p), j(n(p), p)}.$$

Then A_ω and B_ω are lower sets. Furthermore, almost all $p \in D$ belong to an interval of length $2^{\frac{1}{2}-n(p)}$ which is the hypotenuse of a triangle with the square $C_p = C_{n(p), j(n(p), p)}$ in its upper right corner, for which $2W(C_p) > M 2^{n(p)}$. Consequently,

$$W(A_\omega) - W(B_\omega) \geq \sum_p M 2^{-n(p)}/2,$$

where the sum is over those $p \in D$ corresponding to distinct intervals $L_{n(p), j(n(p), p)}$. Since the union of the countably many such intervals is almost all of the diagonal, the sum of $\sum 2^{-n(p)}$ is precisely 1.

Hence $W(A_\omega) - W(B_\omega) \geq \frac{M}{2}$ almost surely, implying that $\max\{|W(A_\omega)|, |W(B_\omega)|\} \geq \frac{M}{4}$. Letting $M \to +\infty$, we see that W is unbounded and therefore discontinuous with probability one over lower sets in $[0,1]^2$.

We conclude that on such $\mathcal{A}$ a strong martingale may be a.s. unbounded and discontinuous. Consequently, without additional restrictions on $\mathcal{A}$ a strong martingale maximal inequality does not exist, since such an inequality immediately implies a.s. convergence of a strong martingale, which then implies monotone outer-continuity of the paths.

The second fundamental example is the set-indexed **Poisson process**.

The Poisson process is a particular case of the notion of point process which will be studied in detail later. The following definition is not new; it has been studied by many authors; see for example Daley and Vere-Jones (1988) or Neveu (1977). Recall that for any increasing process, the index collection can be extended from $\mathcal{A}$ to the Borel sets $\mathcal{B}$ of T.

Definition 3.4.4 *An increasing process* $N = \{N_A, A \in \mathcal{A}\}$ *is called a* Poisson process *if there exists a boundedly finite diffuse Borel measure* $\Lambda(\cdot)$ *such that for every finite family of disjoint bounded Borel sets* $B_1, \ldots, B_k$, *we have:*

$$P\{N_{B_i} = n_i,\ i = 1, \ldots, k\} = \prod_{i=1}^{k} \frac{\Lambda(B_i)^{n_i}}{n_i!}\, e^{-\Lambda(B_i)}.$$

The measure $\Lambda(\cdot)$ *is called the* mean measure *of the process.*

Comments 3.4.5 Although we do not allow fixed atoms in the Poisson process (for any $t \in T$, $P(N_{\{t\}} > 0) = 0$), all of the jump points may live on a subspace of T.

Notice also that the mean measure can be viewed as a (deterministic) increasing process.

Since the Poisson process is increasing, it is clearly both a strong submartingale and a submartingale. As well, it is uniformly integrable on $\mathcal{B}_h$ for every h, and since $N - \Lambda$ is a monotone outer-continuous process with mean 0 independent increments, we have the following proposition which is an immediate consequence of Theorem 3.4.1. In fact, it will be shown in Chapter 5 that the second statement characterizes the Poisson process.

Proposition 3.4.6 *Let N be a Poisson process with mean measure* Λ. *Then $N - \Lambda$ is a strong martingale with respect to its minimal outer-continuous filtration.*

A very important generalization of the Poisson process is the doubly stochastic Poisson process (also called the Cox process). Intuitively, it is a Poisson process such that the mean measure of the process is random. See for example Brémaud (1981), Daley and Vere-Jones (1988), Grandell (1976) or more recently Cojocaru and Merzbach (1999).

Definition 3.4.7 *Let $N = \{N_A, A \in \mathcal{A}\}$ be an adapted increasing process and $\hat{\Lambda} = \{\hat{\Lambda}_A, A \in \mathcal{A}\}$ be an increasing, nonnegative, $\mathcal{G}^*_{\emptyset'}$-*

EXAMPLES 71

measurable process, with $\hat{\Lambda}_{\{t\}} = 0$, such that, for all $C \in \mathcal{C}$, $u \in \mathbf{R}$,

$$E(e^{iuN_C} \mid \mathcal{G}_C^*) = \exp\{(e^{iu} - 1) \cdot \hat{\Lambda}_C\}$$

Then N is called a doubly stochastic $\mathcal{G}^*$-Poisson process, *with mean measure $\hat{\Lambda}$ (or 'directed' by $\hat{\Lambda}$).*

Proposition 3.4.8 *Let N be a doubly stochastic $\mathcal{G}^*$-Poisson process with mean measure $\hat{\Lambda}$. Then $N - \hat{\Lambda}$ is a strong martingale with respect to its minimal filtration and its minimal outer-continuous filtration.*

Proof. For any $C \in \mathcal{C}$,

$$\begin{aligned}
E(N_C - \hat{\Lambda}_C \mid \mathcal{G}_C^*) &= E(N_C \mid \mathcal{G}_C^*) - E(\hat{\Lambda}_C \mid \mathcal{G}_C^*) \\
&= \sum k P(N_C = k \mid \mathcal{G}_C^*) - \hat{\Lambda}_C \\
&= \hat{\Lambda}_C - \hat{\Lambda}_C = 0,
\end{aligned}$$

since $\hat{\Lambda}_C$ is $\mathcal{G}_{\emptyset'}^*$-measurable, and $P(N_C = k \mid \mathcal{G}_C^*) = \exp^{-\hat{\Lambda}(C)} \frac{\hat{\Lambda}(C)^k}{k!}$. $\square$

CHAPTER 4

Decompositions and Quadratic Variation

The Doob-Meyer decomposition of a classical submartingale into the sum of a martingale and a predictable increasing process is central to martingale theory. The increasing process is often referred to as the *compensator* of the submartingale. An important related problem is the existence of a quadratic variation process. This chapter will address the question of both existence and uniqueness of compensators and quadratic variation processes which satisfy a certain condition of predictability analogous to the predictability of the increasing process in the classic Doob-Meyer decomposition. The importance of both the compensator and the quadratic variation process will become apparent in subsequent chapters as we will use both notions to develop martingale characterizations of some well-known processes, and then later to prove weak convergence theorems.

Most of the results in this chapter may be found in Dozzi, Ivanoff and Merzbach (1994) and in Slonowsky (1998) and (1999).

4.1 Definitions

In general, we may define a 'compensator' for a set-indexed submartingale as follows:

Definition 4.1.1 *Let X be a weak (resp. strong) submartingale with $X_{\emptyset'} = 0$. V is said to be a compensator (resp. *-compensator) of X if V is an increasing process and $E[X_C - V_C \mid \mathcal{G}_C] = 0 \ \forall C \in \mathcal{C}$ (resp. $E[X_C - V_C \mid \mathcal{G}_C^*] = 0 \ \forall C \in \mathcal{C}$).*

This definition is very general. Trivially, any increasing process is a (*-)compensator for itself, and there is no requirement for uniqueness or adaptedness, though we note that if the compensator (respectively, *-compensator) is adapted, then $X - V$ is a weak (respectively, strong) martingale. At issue is the appropriate analogue of predictability, as we have defined both the predictable

σ-algebra $\mathcal{P}$, as well as the *-predictable σ-algebra $\mathcal{P}^*$. We shall see that the former is appropriate for weak submartingales, and the latter for strong submartingales.

We now define the quadratic variation process:

Definition 4.1.2 *Let X be an arbitrary square-integrable process. An increasing process Q is called a quadratic variation (respectively, a *-quadratic variation) process for X if for every $C \in \mathcal{C}$*

$$E[(X_C)^2 \mid \mathcal{G}_C] = E[Q_C \mid \mathcal{G}_C]$$

(respectively,

$$E[(X_C)^2 \mid \mathcal{G}_C^*] = E[Q_C \mid \mathcal{G}_C^*]).$$

In this chapter we will show the existence of a unique 'predictable' ('*-predictable') quadratic variation process for a martingale (strong martingale). In the classical theory, the predictable quadratic variation $\langle X \rangle$ of a martingale X is simply the predictable increasing process in the Doob-Meyer decomposition of the submartingale X^2 since $E[X_t^2 - X_s^2 \mid \mathcal{F}_s] = E[(X_t - X_s)^2 \mid \mathcal{F}_s]$. Although a similar approach may be used for the predictable quadratic variation of a martingale, in the case of a set-indexed strong martingale $E[X_C^2 \mid \mathcal{G}_C^*] \neq E[(X_C)^2 \mid \mathcal{G}_C^*]$ in general and so the question of *-predictable quadratic variation must be addressed separately.

The chapter is arranged as follows. In Section 4.2 we prove the existence of admissible measures associated with the various types of processes. Next, we prove the existence of the compensator in the Doob-Meyer decomposition and the quadratic variation process in Section 4.3. We give a second more direct and intuitive proof in Section 4.4. Finally, examples of both compensators and quadratic variation processes for point processes are given in Section 4.5.

Throughout this chapter, it will be assumed that T is metrizable (cf. Theorem 4.2.6) and that all processes X satisfy $X_{\emptyset'} = 0$.

4.2 Admissible Functions and Measures

In this section, we introduce the admissible functions associated with our various types of processes. The key to proving the existence of a predictable compensator or quadratic variation is to extend the admissible function to a σ-additive measure on the appropriate predictable σ-algebra. We begin with the following general definition:

ADMISSIBLE FUNCTIONS AND MEASURES 75

Definition 4.2.1 *The admissible function μ_X associated with an integrable process $X = \{X_A : A \in \mathcal{A}\}$ is defined on the rectangles $\{F \times C : C \in \mathcal{C}, F \in \mathcal{F}\}$, by:*

$$\mu_X(F \times C) = E[I_F X_C].$$

(I_F is the indicator function of F.)

Comments:

1. We notice immediately that for X any integrable and adapted process,
- X is a weak (respectively, strong) martingale if and only if μ_X vanishes on $\mathcal{P}_0$ (respectively, $\mathcal{P}_0^*$).
- X is a weak (respectively, strong) submartingale if and only if $\mu_X \geq 0$ on $\mathcal{P}_0$ (respectively, $\mathcal{P}_0^*$).
- X is an increasing process if and only if X is monotone outer continuous and $\mu_X \geq 0$ on $\{F \times C : C \in \mathcal{C}, F \in \mathcal{F}\}$.

2. Because T is assumed to be σ-compact, it suffices to consider extensions of μ_X on $\Omega \times T'$, where T' is an arbitrary compact subset of T. Thus, without loss of generality, *for the remainder of this chapter we may assume that T is compact and that $T \in \mathcal{A}(u)$.*

3. It is clear that for any integrable process X which is additive on $\mathcal{C}$, μ_X is additive on $\mathcal{F} \times \mathcal{C}$, and hence has an additive extension to the algebra generated by $\mathcal{P}_0$ (resp. $\mathcal{P}_0^*$) which will be non-negative if and only if X is a weak (resp. strong) submartingale. We shall show that for X a weak submartingale, μ_X has a σ-additive extension to $\mathcal{P}$. The analogous result for strong martingales and $\mathcal{P}^*$ is slightly less general, but has a proof which is virtually identical. Additivity of μ_X on $\mathcal{P}_0$ is sufficient to prove the next two lemmas.

Lemma 4.2.2 *Let σ and τ be discrete stopping sets such that $\sigma(\omega) \subseteq \tau(\omega) \ \forall \omega$. Then*

$$\mu_X(\sigma, \tau] = E(X_\tau - X_\sigma).$$

Proof. The proof is almost identical to that of Proposition 3.3.5. If σ and τ are both discrete stopping sets, from Lemma 2.1.5 we may assume that there exist finitely many disjoint sets $C_1, \ldots, C_k \in \mathcal{C}$ such that $(\emptyset, \tau] = \cup_{j=1}^k F_j \times C_j$, $(\emptyset, \sigma] = \cup_{j=1}^k G_j \times C_j$, where $F_j, G_j \in \mathcal{G}_{C_j}$ and $G_j \subseteq F_j, j = 1, \ldots, k$. Then

$$\begin{aligned}(\sigma, \tau] &= (\cup_{j=1}^k F_j \times C_j) \setminus (\cup_{j=1}^k G_j \times C_j) \\ &= \cup_{j=1}^k (F_j \times C_j) \setminus (G_j \times C_j) \text{ since the } C_j\text{'s are disjoint} \\ &= \cup_{j=1}^k (F_j \setminus G_j) \times C_j.\end{aligned}$$

Thus,

$$\mu_X(\sigma,\tau] = E\left[\sum_{j=1}^{k} I(F_j \setminus G_j) X_{C_j}\right]$$

$$= E\left[\sum_{j=1}^{k} I(F_j) X_{C_j}\right] - E\left[\sum_{j=1}^{k} I(G_j) X_{C_j}\right]$$

$$= E(X_\tau - X_\sigma).$$

□

Before continuing with the extension of the admissible function, we introduce the analogue which is associated with the squares of increments of a process:

Definition 4.2.3 *The admissible square function $\mu_X^{(2)}$ associated with a square integrable process $X = \{X_A : A \in \mathcal{A}\}$ is defined on the rectangles $\{F \times C : C \in \mathcal{C}, F \in \mathcal{F}\}$, by:*

$$\mu_X^{(2)}(F \times C) = E[I_F(X_C)^2].$$

Note: The admissible square function $\mu_X^{(2)}$ is not necessarily the admissible function of the process X^2 since in general $E[I_F(X_C)^2] \neq E[I_F X_C^2]$. However, we do have the following:

Lemma 4.2.4 *(i) If X is a square integrable martingale and if SHAPE and (CI) hold, then $\mu_X^{(2)}$ is finitely additive on $\mathcal{P}_0$.*
(ii) If X is a square integrable strong martingale, then $\mu_X^{(2)}$ is finitely additive on $\mathcal{P}_0^*$.

Proof. (i) If (CI) holds and X is a martingale, then by Proposition 3.1.8, X^2 is a weak submartingale whose admissible measure is nonnegative and additive on $\mathcal{P}_0$. It remains to observe that $\mu_{X^2} = \mu_X^{(2)}$ on $\mathcal{P}_0$ since by Lemma 3.1.2 and Proposition 3.1.7, $E[I_F(X_C)^2] = E[I_F X_C^2]$ for every $C \in \mathcal{C}$ and $F \in \mathcal{G}_C$.
(ii) For a strong martingale X, it is not true that X^2 is a strong submartingale. However, it is enough (cf. Slonowsky (1998), Gushchin (1982)) to show that $\mu_X^{(2)}$ is additive on sets of the form $F \times C = F \times (\cup_{i=1}^n C_i)$, where $F \in \mathcal{G}_C^*$ and $C = (\cup_{i=1}^n C_i)$ is a disjoint union. We have

$$\mu_X^{(2)}(F \times C) = E\left[I_F(\sum_{i=1}^{n} X_{C_i})^2\right]$$

ADMISSIBLE FUNCTIONS AND MEASURES 77

$$= E\left[I_F \sum_{i=1}^{n}(X_{C_i})^2\right] + E\left[I_F \sum_{i \neq j} X_{C_i} X_{C_j}\right].$$

We are done if it can be shown that $E[I_F X_{C_i} X_{C_j}] = 0$ if $i \neq j$. By considering the finite sub-semilattice $\mathcal{A}'$ generated by the sets in $\mathcal{A}$ which define C_i and C_j, it is easily seen that there exist disjoint left-neighbourhoods $D_1, ..., D_k \in \mathcal{C}^{\ell}(\mathcal{A}')$ such that $C_i = \cup_{r=1}^{h} D_r$ and $C_j = \cup_{s=h+1}^{k} D_s$. By Proposition 1.4.6, if $r \neq s$ then either D_r is $\mathcal{G}_{D_s}^*$-measurable or D_s is $\mathcal{G}_{D_r}^*$-measurable. In either case, since $D_r, D_s \subseteq C$, $F \in \mathcal{G}_{D_r}^*$ and $F \in \mathcal{G}_{D_s}^*$, so by the strong martingale property $E[I_F X_{D_r} X_{D_s}] = 0$. Therefore,

$$E[I_F X_{C_i} X_{C_j}] = \sum_{r=1}^{h} \sum_{s=h+1}^{k} E[I_F X_{D_r} X_{D_s}] = 0.$$

This completes the proof. $\square$

Before considering the extension of the admissible functions to measures, we will need one technical lemma:

Lemma 4.2.5 *(i) (Dozzi, Ivanoff and Merzbach (1994)) If X is an L^1- monotone outer-continuous (weak or strong) submartingale, then for any set $C \in \mathcal{C}$, there exists an increasing sequence of sets (C_n) in $\mathcal{C}$ such that $\cup_{n=1}^{\infty} C_n = C$, $\overline{C_n} \subseteq C$ and $\lim_{n \to \infty} E[X_{C \setminus C_n}] = 0$.*
(ii) (Slonowsky (1998)) If X is an L^2 strong martingale which is L^2-mononotone outer-continuous, then for any set $C \in \mathcal{C}$, there exists an increasing sequence of sets (C_n) in $\mathcal{C}$ such that $\cup_{n=1}^{\infty} C_n = C$, $\overline{C_n} \subseteq C$ and $\lim_{n \to \infty} E[(X_{C \setminus C_n})^2] = 0$.

Proof. Let $C = A \setminus \cup_{i=1}^{k} A_i$ be a maximal representation of C in $\mathcal{A}$ and let $C_n = A \setminus \cup_{i=1}^{k} g_n(A_i)$. Clearly

$$\overline{C_n} = A \setminus (\cup_{i=1}^{k} g_n(A_i))^{\circ} \subseteq A \setminus \cup_{i=1}^{k} (g_n(A_i))^{\circ} \subseteq C.$$

Now, for any subset $\mathcal{I} \subseteq \{1, ..., k\}$, the sequence of sets

$$A \cap \cap_{i \in \mathcal{I}} g_n(A_i) = B_{\mathcal{I}}^n$$

decreases to $B_{\mathcal{I}} = \cap_{i \in \mathcal{I}} A_i$. For (i), by L^1 outer-continuity, $X_{B_{\mathcal{I}}^n} \to X_{B_{\mathcal{I}}}$ in L^1 norm as $n \to \infty$, so $E[X_{C_n}] \to_{n \to \infty} E[X_C]$. The proof of (ii) is analogous. $\square$

We now turn to the question of extending the admissible functions μ_X and $\mu_X^{(2)}$ of the various types of processes to σ-additive admissible measures (also denoted by μ_X and $\mu_X^{(2)}$) on appropriate σ-algebras. The easiest case is the admissible measure of an increasing process.

Theorem 4.2.6 *If X is an increasing process, then μ_X can be extended uniquely to a σ-additive measure on the product space $(\Omega \times T, \mathcal{F} \times \mathcal{B})$.*

Proof. For each $C \in \mathcal{C}$ fixed, $\mu(F, C) = \mu_X(F \times C)$ is a σ-additive positive measure on $\mathcal{F}$, and for each $F \in \mathcal{F}$ fixed, $\mu(F, C) = \mu_X(F \times C)$ is a σ-additive positive measure in $\mathcal{B}$ (by the L^1 outer-continuity of V). Therefore, μ is a positive bimeasure. Since T is metrizable, a result of Horowitz (1977) ensures that μ can be extended uniquely to a measure on the product space. □

Next, we define various types of 'class D' processes, one of which was introduced in the preceding chapter. Recall that τ^f denotes the family of discrete stopping sets; that is, stopping sets taking on finitely many configurations.

Definition 4.2.7 *Let X be an adapted process on $\mathcal{A}$. X is said to be of:*

- *class D if the family $\{X_\xi : \xi \in \tau^f\}$ is uniformly integrable,*

- *class D' if the family*

$$\{\sum_{C \in \mathcal{C}^t(\mathcal{A}')} E[X_C \mid \mathcal{G}_C] : \mathcal{A}' \text{ a finite sub-semilattice of } \mathcal{A}\}$$

is uniformly integrable,

- *class $(D')^*$ if the family*

$$\{\sum_{C \in \mathcal{C}^t(\mathcal{A}')} E[X_C \mid \mathcal{G}_C^*] : \mathcal{A}' \text{ a finite sub-semilattice of } \mathcal{A}\}$$

is uniformly integrable.

Comments:
1. We will be proving the existence of a Doob-Meyer decomposition for submartingales of class D or D'. Although the two classes are equivalent for submartingales indexed by $\mathbf{R}_+$, it is not clear if this is true in our more general setting.

2. We do not have a class D in the context of strong submartingales, as we have found no suitable analogue of Proposition 2.1.6 which characterizes stochastic intervals in terms of predictable sets.

Theorem 4.2.8 *(Dozzi, Ivanoff and Merzbach (1994))* μ_X *can be extended uniquely to a σ-additive measure on $\mathcal{P}$ if X is an L^1-monotone outer-continuous weak submartingale of class D or D'.*

Proof. We follow the classic sort of argument here; see for example §8.4 of Métivier and Pellaumail (1980).

Since μ_X is finite on $\Omega \times T$, and finitely additive on $\mathcal{P}_0(u)$, which is the algebra generated by $\mathcal{P}_0$, it is enough to show that for any sequence (R_n) of sets in $\mathcal{P}_0(u)$ such that $R_n \downarrow \emptyset$, $\mu_X(R_n) \downarrow 0$.

We shall assume that $\mu_X(R_n) \downarrow 3a > 0$, and show that this leads to a contradiction.

Let $\{B_k^n; 1 \leq k \leq b(n)\}$ be a finite partition of R_n, where $B_k^n = F_k^n \times C_k^n \in \mathcal{P}_0$ for all n, k. By Proposition 4.2.5 we may choose a set E_k^n such that $\overline{E_k^n} \subseteq C_k^n$, and

$$\mu_X(\Omega \times C_k^n \backslash E_k^n) = E(X_{C_k^n \backslash E_k^n}) \leq a/nb(n). \tag{4.1}$$

Let

$$\begin{aligned}
E_n &:= \cup_{k=1}^{b(n)} F_k^n \times E_k^n \\
\tilde{E}_n &:= \cup_{k=1}^{b(n)} F_k^n \times \overline{E_k^n} \\
D_n &:= \cap_{k=1}^{n} E_k \\
\tilde{D}_n &:= \cap_{k=1}^{n} \tilde{E}_k \\
S_n &:= \cup_{k=1}^{b(n)} (\Omega \times \{C_k^n \backslash E_k^n\})
\end{aligned}$$

Note: E_n, D_n, S_n are in $\mathcal{P}_0(u)$.

By definition, it is straightforward to see that $R_n \subseteq D_n \cup (\cup_{k=1}^n S_k)$. Since $D_n, S_n \in \mathcal{P}_0(u)$ for all n, by (4.1) we have

$$\mu_X(R_n) \leq \mu_X(D_n) + \sum_{k=1}^{n} \mu_X(S_k) \leq \mu_X(D_n) + a.$$

We now observe that $\tilde{E}_n \subseteq R_n$, and so $\tilde{D}_n \downarrow \emptyset$. Fix $\omega \in \Omega$. Since $\tilde{D}_n(\omega)$ is compact and decreasing in n, there exists $n(\omega) < \infty$ such that $\tilde{D}_n(\omega) = \emptyset$, for all $n \geq n(\omega)$. Thus, $P\{\omega : D_n(\omega) \neq \emptyset\} \to 0$.

Suppose first that X is of class D. Let ξ_n be the début of D_n. Since $D_n(\omega) = \emptyset \Rightarrow \xi_n(\omega) = T$, it follows that $P\{\xi_n \neq T\} \to 0$. Since $D_n \subseteq (\xi_n, T]$, we have

$$\begin{aligned}
\mu_X(D_n) &\leq \mu_X(\xi_n, T] = E(X_T - X_{\xi_n}) \\
&= \int_{\{\xi_n \neq T\}} X_T dP - \int_{\{\xi_n \neq T\}} X_{\xi_n} dP \to 0,
\end{aligned}$$

by uniform integrability.

Finally, if X is of class D', and $D_n = \cup_{i=1}^{k(n)} F_i^n \times C_i^n$ ($C_1^n, \ldots, C_{k(n)}^n$ are disjoint), then without loss of generality it may be assumed that the sets D_n are the left-neighbourhoods of some finite sub-semilattice $\mathcal{A}'$ of $\mathcal{A}$.

$$\begin{aligned} 0 \leq \mu_X(D_n) &= \sum_i E(I_{F_i^n} X_{C_i^n}) \\ &= \sum_i \int_{F_i^n} E(X_{C_i^n} | \mathcal{G}_{C_i^n}) \, dP \\ &\leq \int_{\cup_i F_i^n} \sum_{i=1}^{k(n)} |E(X_{C_i^n} | \mathcal{G}_{C_i^n})| \, dP. \end{aligned}$$

But $\cup_{i=1}^{k(n)} F_i^n = \{\omega : D_n(\omega) \neq \emptyset\}$. By uniform integrability, there exists $\delta > 0$ such that $P(\Omega') < \delta$ implies that, for all n, $\int_{\Omega'} \sum_{i=1}^{k(n)} |E(X_{C_i^n}|\mathcal{G}_{C_i^n})| dP < a$. Thus, since $P\{\omega : D_n(\omega) \neq \emptyset\} \to 0$, there exists n_0 such that $n > n_0 \Rightarrow \mu_X(D_n) < a$.

Therefore, $\mu_X(R(n)) \leq 2a$ for all n sufficiently large. This contradicts the assumption that $a > 0$, and the theorem is proved. $\square$

For strong submartingales, we have the following from Slonowsky (1998):

Theorem 4.2.9 μ_X can be extended uniquely to a σ-additive measure on $\mathcal{P}^*$ if X is an L^1-monotone outer-continuous strong submartingale of class $(D')^*$.

Proof. The proof follows exactly along the lines of that of the preceding theorem for class D' weak submartingales, replacing $\mathcal{P}_0$ by $\mathcal{P}_0^*$. $\square$

Finally, for martingales and strong martingales, we have:

Theorem 4.2.10 *(i)* $\mu_X^{(2)}$ can be extended uniquely to a σ-additive measure on $\mathcal{P}$ if (CI) and SHAPE hold and if X is an L^2-monotone outer-continuous martingale such that X^2 is of class D or D'.
(ii) $\mu_X^{(2)}$ can be extended uniquely to a σ-additive measure on $\mathcal{P}^*$ if X is an L^2-monotone outer-continuous strong martingale and if the class

$$\{ \sum_{C \in \mathcal{C}^\ell(\mathcal{A}')} E[(X_C)^2 \mid \mathcal{G}_C^*] : \mathcal{A}' \text{ a finite sub-semilattice of } \mathcal{A}\}$$

is uniformly integrable.

Proof. (i) As noted before, in this case $E[X_C^2 \mid \mathcal{G}_C] = E[(X_C)^2 \mid \mathcal{G}_C]$, and so X^2 is a L^1-monotone outer-continuous weak submartingale and $\mu_X^{(2)} = \mu_{X^2}$. The result follows from Theorem 4.2.8.
(ii) The proof again mimics the proof of Theorem 4.2.8 for the case of a class D' weak submartingale, replacing $\mathcal{G}_{C_i^n}$ with $\mathcal{G}_{C_i^n}^*$, μ_X with $\mu_X^{(2)}$ and X_C with $(X_C)^2$ for all $C \in \mathcal{C}$. □

4.3 The Doob-Meyer Decomposition and the Quadratic Variation Process

In this section, our main result is a sort of converse to Theorem 4.2.6. Generally speaking, we will show that for any measure μ on $\mathcal{P}$ (respectively, $\mathcal{P}^*$), there exists an increasing process whose admissible measure coincides with μ on $\mathcal{P}$ (respectively, $\mathcal{P}^*$). In addition, if the increasing process is 'predictable' in a sense which will be made precise below, it is unique (see Theorem 4.3.6 below). This single theorem allows us to prove the existence of both the (*-)compensator in the Doob-Meyer decomposition and the quadratic variation process. The results for martingales and weak submartingales are from Dozzi, Ivanoff and Merzbach (1994). The results for strong martingales and submartingales were proven in Slonowsky (1998) and (1999) using a different method. This approach will be explored in the next section.

In what follows, it must be assumed that :

Assumption 4.3.1 *Any function which is additive on $\mathcal{A}$ may be extended to a function which is additive on $\mathcal{C}$.*

It has already been pointed out that this is the case if SHAPE holds. It is also straightfoward to show (cf. Slonowsky 1998) that an additive extension exists if whenever $A \setminus \cup_{i=1}^k A_i = A' \setminus \cup_{j=1}^{k'} A'_j$, then there exists $N \in \mathbf{N}$ such that

$$g_n(A) \setminus \cup_{i=1}^k g_n(A_i) = g_n(A') \setminus \cup_{j=1}^{k'} g_n(A'_j)$$

for each $n \geq N$. We note that all of our examples satisfy Assumption 4.3.1.

The 'theorem of predictable projection' of the classical theory is used in order to obtain a predictable process in the Doob-Meyer decomposition. In the multiparameter case, the existence of a (*)-predictable projection is a result of (CI). In our general setting, we must assume that an appropriate projection exists using an approach introduced by Allain (1983). First we consider $\mathcal{P}$.

Assumption 4.3.2 *There exists a T-indexed filtration $\{\mathcal{F}_t^-, t \in T\}$ such that for any $C \in \mathcal{C}$ and $A \in \mathcal{A}$, $t \in C \cap A$ implies $\mathcal{G}_C \subseteq \mathcal{F}_t^- \subseteq \mathcal{F}_A$. Moreover, for all $F \in \mathcal{F}$, there exists a unique (up to indistinguishability) T-indexed $\mathcal{P}$-measurable process $Y = \{Y_t, t \in T\}$ such that $Y_t = E[I_F \mid \mathcal{F}_t^-]$.*

Next we consider $\mathcal{P}^*$.

Assumption 4.3.3 *There exists a T-indexed filtration $\{(\mathcal{F}_t^*)^-, t \in T\}$ such that for any $C \in \mathcal{C}$, $t \in C$ implies $\mathcal{G}_C^* \subseteq (\mathcal{F}_t^*)^-$. Moreover, for all $F \in \mathcal{F}$, there exists a unique (up to indistinguishability) T-indexed $\mathcal{P}^*$-measurable process $Y = \{Y_t, t \in T\}$ such that $Y_t = E[I_F \mid (\mathcal{F}_t^*)^-]$.*

Both assumptions appear very natural and we conjecture that they are also necessary for the respective Doob-Meyer decompositions. Specific constructions of the filtrations $\{\mathcal{F}_t^-, t \in T\}$ and $\{(\mathcal{F}_t^*)^-, t \in T\}$ are given in the next section. We note that any T-indexed $\mathcal{P}$-measurable process Y is $\{\mathcal{F}_A\}$-adapted in the sense that Y_t is $\mathcal{F}_A$-measurable for any A such that $t \in A$. The same is not necessarily true for T-indexed $\mathcal{P}^*$-measurable processes.

We now define what is meant by a 'predictable' $\mathcal{A}$-indexed increasing process. Since the index set of the process is $\mathcal{A}$ and not T, we cannot simply require such a process to be $\mathcal{P}$- or $\mathcal{P}^*$- measurable. Instead we extend the idea of the dual predictable projection which is equivalent in the classical theory, and we have the following two definitions.

Definition 4.3.4 *An increasing process $V = \{V_A, A \in \mathcal{A}\}$ is predictable if for any $F \in \mathcal{F}$ and $C \in \mathcal{C}$,*

$$\mu_V(F \times C) = E[I_F V_C] = \int E[I_F \mid \mathcal{F}_t^-](\omega) I_C(t) d\mu_V.$$

Definition 4.3.5 *An increasing process $V = \{V_A, A \in \mathcal{A}\}$ is $*$-predictable if for any $F \in \mathcal{F}$ and $C \in \mathcal{C}$,*

$$\mu_V(F \times C) = E[I_F V_C] = \int E[I_F \mid (\mathcal{F}_t^*)^-](\omega) I_C(t) d\mu_V.$$

The existence of ($*$-)predictable compensators and quadratic variation processes will follow directly from the next theorem.

Theorem 4.3.6 *(i) Suppose that Assumptions 4.3.1 and 4.3.2 hold*

COMPENSATOR AND QUADRATIC VARIATION 83

and let μ be any measure on $\mathcal{P}$. Then there exists a unique (up to indistinguishability) increasing adapted predictable process V such that $\mu_V = \mu$ on $\mathcal{P}$.

(ii) Suppose that Assumptions 4.3.1 and 4.3.3 hold and let μ be any measure on $\mathcal{P}^*$. Then there exists a unique (up to indistinguishability) increasing *-predictable process V such that $\mu_V = \mu$ on $\mathcal{P}^*$.

Proof. We begin with the proof of (i):
For $F \in \mathcal{F}$ and $C \in \mathcal{C}$ we define

$$\nu_C(F) = \int_{\Omega \times T} E[I_F \mid \mathcal{F}_t^-](\omega) I_C(t) d\mu.$$

For fixed C, $\nu_C(\cdot)$ is σ-additive and dominated by P, so we may define $\nu_C' := d\nu_C/dP$; $\nu'_{(\cdot)}$ is an integrable process which is monotone outer-continuous in mean.

Let $\mathcal{C}_0 = \cup_n \mathcal{C}^\ell(\mathcal{A}_n)$. Since $\mathcal{C}_0$ is countable, there exists an event $\Omega_0 \subseteq \Omega$ such that $P(\Omega_0) = 1$, $\nu'_{(\cdot)}$ is additive for all $\omega \in \Omega_0$ and $\nu'_C(\omega) \geq 0$ for all $\omega \in \Omega_0$ and all $C \in \mathcal{C}_0$.

We define V as follows:

$$V_A = \lim_{n \to \infty} \nu'_{g_n(A)} I_{\Omega_0}.$$

Since we may assume that $\mathcal{A}_n$ satisfies SHAPE and since g_n preserves unions for each n, it is straightforward that the sample paths of V are additive on $\mathcal{A}$, and hence V can be extended to a process with additive sample paths on $\mathcal{C}$. We now verify that V satisfies the required properties.

(a) *V is an increasing process*:
If $C = A \setminus B \in \mathcal{C}, A \in \mathcal{A}, B \in \mathcal{A}(u)$, then

$$V_C = V_A - V_B = (\lim_n \nu'(g_n(A)) - \lim_n \nu'(g_n(B))) I_{\Omega_0} \geq 0.$$

Also, if $A_n \downarrow A$, $A_n, A \in \mathcal{A}$, then

$$\begin{aligned}
\lim_n V(A_n) &= \lim_n \lim_m \nu'_{g_m(A_n)} I_{\Omega_0} \\
&= \inf_n \inf_m \nu'_{g_m(A_n)} I_{\Omega_0} \\
&= \inf_m \inf_n \nu'_{g_m(A_n)} I_{\Omega_0},
\end{aligned}$$

since all sequences are nonincreasing. But since g_m preserves intersections and $\mathcal{A}_m$ is finite, $g_m(A_n) = g_m(A)$ for all n sufficiently large. Therefore,

$$\lim_n V(A_n) = \lim_m \nu'_{g_m(A)} I_{\Omega_0} = V(A),$$

and V is a monotone outer-continuous modification of ν'.
(b) *V is adapted*:

We now note that for any $F \in \mathcal{F}$, $E[I_F V_C] = \nu_C(F)$ and

$$\int_\Omega I_F(\omega) d\nu_C(\omega) = \nu_C(F) = \int_{\Omega \times T} E[I_F \mid \mathcal{F}_t^-](\omega) I_C(t) d\mu. \quad (4.2)$$

By a classical argument, this implies for any $Y \in L^\infty(\Omega, \mathcal{F}, P)$ that

$$\int_\Omega Y(\omega) d\nu_C(\omega) = \int_{\Omega \times T} E[Y \mid \mathcal{F}_t^-] I_C(t) d\mu. \quad (4.3)$$

Next we observe that V is an adapted process since for any $F \in \mathcal{F}$,

$$\begin{aligned}
\nu_A(F) &= \int_{\Omega \times T} E[I_F \mid \mathcal{F}_t^-] I_A(t) d\mu \\
&= \int_{\Omega \times T} E[E(I_F \mid \mathcal{F}_A) \mid \mathcal{F}_t^-] I_A(t) d\mu \\
&= \int_\Omega E[I_F \mid \mathcal{F}_A] d\nu_A(\omega)
\end{aligned}$$

where the last equality follows from (4.3).

(c) $\mu_V = \mu$ on $\mathcal{P}$:

Note that $\mu_V = \mu$ on $\mathcal{P}$ since $t \in C \in \mathcal{C}$ implies $\mathcal{G}_C \subseteq \mathcal{F}_t^-$ and so for $F \times C$ in $\mathcal{P}_0$,

$$\begin{aligned}
\mu_V(F \times C) &= E[I_F V_C] = \int_{\Omega \times T} E[I_F \mid \mathcal{F}_t^-](\omega) I_C(t) d\mu \text{ by (4.2)} \\
&= \int_{\Omega \times T} I_F(\omega) I_C(t) d\mu \text{ since } F \in \mathcal{G}_C \subseteq \mathcal{F}_t^- \\
&= \mu(F \times C).
\end{aligned}$$

(d) *V is predictable*:

Predictability of V then follows from the construction: for $F \in \mathcal{F}$ and $C \in \mathcal{C}$,

$$\begin{aligned}
\mu_V(F \times C) &= E[I_F V_C] = \nu_C(F) \\
&= \int_{\Omega \times T} E[I_F \mid \mathcal{F}_t^-](\omega) I_C(t) d\mu \\
&= \int_{\Omega \times T} E[I_F \mid \mathcal{F}_t^-](\omega) I_C(t) d\mu_V
\end{aligned}$$

since $\mu_V = \mu$ on $\mathcal{P}$.

(e) *V is unique*:

We have shown the existence of a monotone outer-continuous

adapted increasing process V such that $\mu_V = \mu$ on $\mathcal{P}$. If two such processes V and V' exist, then for $F \in \mathcal{F}$ and $C \in \mathcal{C}$,

$$\mu_V(F \times C) = \int_{\Omega \times T} E[I_F \mid \mathcal{F}_t^-](\omega) I_C(t) d\mu_V$$
$$= \int_{\Omega \times T} E[I_F \mid \mathcal{F}_t^-](\omega) I_C(t) d\mu_{V'} = \mu_{V'}(F \times C).$$

Therefore, $E[I_F V_C] = E[I_F V_C']$ for every $F \in \mathcal{F}$, and so $V = V'$ up to a modification. Monotone outer-continuity then ensures that V' and V are indistinguishable.

This completes the proof of (i).

The proof of (ii) is completely analogous, replacing $\mathcal{F}_t^-$ with $(\mathcal{F}_t^*)^-$, $\mathcal{G}_C$ with $\mathcal{G}_C^*$ and $\mathcal{P}$ with $\mathcal{P}^*$ in steps (a), (c), (d) and (e). However, the process V will no longer necessarily be adapted. □

The Doob-Meyer decompositions below now follow as corollaries of Theorems 4.2.6, 4.2.8, 4.2.9 and 4.3.6.

Corollary 4.3.7 Doob-Meyer Decompositions

(i) Suppose that Assumptions 4.3.1 and 4.3.2 hold and let X be a weak submartingale satisfying one of the following properties:

- *X is an increasing process.*
- *X is L^1-monotone outer-continuous of class D or D'.*

Then X has a unique (up to indistinguishability) predictable compensator V and $X - V$ is a weak martingale.

(ii) Suppose that Assumptions 4.3.1 and 4.3.3 hold and let X be a strong submartingale satisfying one of the following properties:

- *X is an increasing process.*
- *X is L^1-monotone outer-continuous strong submartingale of class $(D')^*$.*

*Then X has a unique (up to indistinguishability) *-predictable *-compensator V. $X - V$ is a strong martingale if V is adapted.*

Comment: Suppose that Assumption 4.3.2 holds and that $T \in \mathcal{A}$. Then every positive (strong) submartingale X has a unique Doob-Meyer decomposition $X = M + V$, where M is a weak martingale and V is a predictable increasing process. This follows since if $T \in \mathcal{A}$, X is of class D, by Corollary 3.3.2.

Sufficient conditions for the existence of quadratic variation processes are given in the next corollary. Part (i) is included in Corollary 4.3.7(i). Part (ii) follows from Theorems 4.2.10 and 4.3.6, and appears in Slonowsky (1998) and (1999).

Corollary 4.3.8 *(i) If Assumptions 4.3.1, 4.3.2, (CI) and SHAPE hold and if X is an L^2-monotone outer-continuous martingale such that X^2 is of class D or D', then X has a unique (up to indistinguishability) predictable quadratic variation process.*
(ii) If Assumptions 4.3.1 and 4.3.3 hold and if X is an L^2-monotone outer-continuous strong martingale such that the class

$$\{ \sum_{C \in \mathcal{C}^\ell(\mathcal{A}')} E[(X_C)^2 \mid \mathcal{G}_C^*] : \mathcal{A}' \text{ a finite sub-semilattice of } \mathcal{A}\}$$

is uniformly integrable, then X has a unique (up to indistinguishability) *-predictable *-quadratic variation process.

The following theorem identifies the quadratic variation of the set-indexed Brownian motion defined in the preceding chapter. Its proof is virtually identical to that of Proposition 3.4.1 and so will be omitted.

Theorem 4.3.9 *Let B be a monotone outer-continuous Brownian motion process on $\mathcal{A}$ with variance measure Λ. Λ is both a quadratic variation and a *-quadratic variation process for B with respect to its minimal filtration and its minimal outer-continuous filtration. Λ is trivially adapted, predictable and *-predictable.*

As noted in Chapter 3, the Brownian motion process is not necessarily monotone outer-continuous. In this case, the preceding theorem remains true only for the minimal filtration $\mathcal{F}^0$.

The compensator of the Poisson process and the quadratic variation of the resulting martingale will be discussed in Section 4.5.

4.4 Discrete Approximations

A more intuitive approach to predictable compensators and quadratic variation processes is through discrete approximations. For this, we use the sublattices $\mathcal{A}_n$ and the classes $\mathcal{C}^\ell(\mathcal{A}_n)$ of left-neighbourhoods generated by them, and we shall continue to assume that $T \in \mathcal{A}(u)$. We note that $\mathcal{C}^\ell(\mathcal{A}_n)$ is a partition of T, and for $t \in T$, we let C_t^n be that set in $\mathcal{C}^\ell(\mathcal{A}_n)$ to which t belongs (i.e. $C_t^n = C \in \mathcal{C}^\ell(\mathcal{A}_n)$ if and only if $t \in C$).

We now restrict the filtrations in Assumptions 4.3.2 and 4.3.3 by explicitly defining the σ-algebras $\mathcal{F}_t^-$ and $(\mathcal{F}_t^*)^-$ in the obvious way:

Assumption 4.4.1 *For each t, let $\mathcal{H}_t = \vee_n \mathcal{G}_{C_t^n}$. Then for each*

$F \in \mathcal{F}$ there exists a unique (up to indistinguishability) $\mathcal{P}$-measurable process $Y_t = E[I_F \mid \mathcal{H}_t]$.

Assumption 4.4.2 For each t, let $\mathcal{H}_t^* = \vee_n \mathcal{G}_{C_t^n}^*$. Then for each $F \in \mathcal{F}$ there exists a unique (up to indistinguishability) $\mathcal{P}^*$-measurable process $Y_t = E[I_F \mid \mathcal{H}_t^*]$.

It is easy to see using separability from above that Assumption 4.4.1 $\Rightarrow$ Assumption 4.3.2 and that Assumption 4.4.2 $\Rightarrow$ Assumption 4.3.3.

We shall prove the existence of a discrete weak L^1 approximation of the compensator of a strong submartingale. The proofs for the approximations of the compensator of a weak submartingale and of the *-predictable quadratic variation are similar, using (respectively) $\mathcal{P}$ instead of $\mathcal{P}^*$ and X_C^2 instead of X_C in the proof. Details for the weak submartingale case are given in Dozzi, Ivanoff and Merzbach (1994), and for the *-predictable quadratic variation in Slonowsky (1998).

Our construction allows us to prove that in certain circumstances, the *-predictable compensator (respectively, quadratic variation) is adapted. For this, we will need the following assumption:

Assumption 4.4.3 Let $B \in \mathcal{A}(u)$, $C \in \mathcal{C}$, $C \subseteq B$. Let $C = A \setminus \cup_1^k A_i$ be a maximal representation of C in B. Then $A_i \cap A_j \subseteq A$, $\forall 1 \leq i, j \leq k$.

We begin with a *discrete* process. Recall that $\mathcal{C}_n$ denotes the class of sets in $\mathcal{C}$ of the form $\{A \setminus B, A \in \mathcal{A}_n, B \in \mathcal{A}_n(u)\}$. Note that each set $C \in \mathcal{C}_n$ has a unique representation $C = \cup_{i=1}^k C_i$, where $C_1, ..., C_k \in \mathcal{C}^\ell(\mathcal{A}_n)$.

Definition 4.4.4 Fix n. Let X be an integrable and adapted process indexed by $\mathcal{A}_n$ which has an additive extension to $\mathcal{C}_n$.

1. X is $\mathcal{A}_n$-increasing if for any $C \in \mathcal{C}_n$, $X_C \geq 0$ a.s.
2. X is an $\mathcal{A}_n$-martingale if for any $A, B \in \mathcal{A}_n, A \subseteq B$, then $E[X_B \mid \mathcal{F}_A] = X_A$.
3. X is a weak (resp. strong) $\mathcal{A}_n$-(sub)martingale if for any $C \in \mathcal{C}_n$, $E[X_C \mid \mathcal{G}_C] = (\geq)0$ (resp. $E[X_C \mid \mathcal{G}_C^*] = (\geq)0$).

Lemma 4.4.5 (i) Every weak $\mathcal{A}_n$-submartingale X can be decomposed into a sum of a weak $\mathcal{A}_n$-martingale and an $\mathcal{A}_n$-increasing process V.
(ii) If X is a strong $\mathcal{A}_n$-submartingale then there exists an $\mathcal{A}_n$-increasing process V such that for every $C \in \mathcal{C}_n$, $E[X_C \mid \mathcal{G}_C^*] =$

$E[V_C \mid \mathcal{G}_C^*]$. If Assumption 4.4.3 and the conditions of Lemma 1.4.3 hold, then $X - V$ is a strong $\mathcal{A}_n$-martingale.

Proof. We shall prove the more difficult result,(ii):
Let $\overline{V}_C = E[X_C \mid \mathcal{G}_C^*]$ for all $C \in \mathcal{C}^\ell(\mathcal{A}_n)$. Define V on $\mathcal{A}_n$ as follows: for $A \in \mathcal{A}_n$,

$$V_A = \sum_{C \subseteq A, C \in \mathcal{C}^\ell(\mathcal{A}_n)} \overline{V}_C.$$

A tedious but straightforward calculation shows that $V_C = \overline{V}_C$ for all $C \in \mathcal{C}^\ell(\mathcal{A}_n)$ and that V is additive on $\mathcal{C}_n$. Therefore, V is $\mathcal{A}_n$-increasing and it is straightforward to show that V is adapted if the conditions of Lemma 1.4.3 hold. Finally, if $C = \cup_{i=1}^k C_i$, $C \in \mathcal{C}_n$, $C_1, ..., C_k \in \mathcal{C}^\ell(\mathcal{A}_n)$,

$$E[X_C \mid \mathcal{G}_C^*] = \sum_{i=1}^k E[E[X_{C_i} \mid \mathcal{G}_{C_i}^*] \mid \mathcal{G}_C^*]$$

$$= \sum_{i=1}^k E[E[V_{C_i} \mid \mathcal{G}_{C_i}^*] \mid \mathcal{G}_C^*] = E[V_C \mid \mathcal{G}_C^*].$$

Thus, V is the desired process, and $X - V$ is a strong $\mathcal{A}_n$-martingale if V is adapted. □

Analogously, we have

Lemma 4.4.6 *If X is a square-integrable strong $\mathcal{A}_n$-martingale then there exists an $\mathcal{A}_n$-increasing process Q such that $E[(X_C)^2 \mid \mathcal{G}_C^*] = E[Q_C \mid \mathcal{G}_C^*]$. If Assumption 4.4.3 and the conditions of Lemma 1.4.3 hold, then Q is $\mathcal{A}_n$-adapted.*

Now, if X is a weak (resp. strong) $\mathcal{A}$-indexed submartingale, then $X^n = \{X_A : A \in \mathcal{A}_n\}$ is a weak (resp. strong) $\mathcal{A}_n$-martingale. Denote by V^n the $\mathcal{A}_n$-increasing process of X^n as defined in Lemma 4.4.5(i) (resp. (ii)). For X a strong martingale, let Q^n be the $\mathcal{A}_n$-increasing process defined in Lemma 4.4.6. We have the following:

Theorem 4.4.7 *(i) Assume that Assumptions 4.3.1 and 4.4.1 hold and that X is a weak submartingale satisfying one of the two properties stated in Corollary 4.3.7 (i). Then for any $A \in \mathcal{A}$,*

$$V_A = \lim_{m \to \infty} \lim_{n \to \infty, n \geq m} V_{g_m(A)}^n$$

in the weak L^1 topology, where V is the predictable compensator of X.

(ii) Assume that Assumptions 4.3.1 and 4.4.2 hold and that X is a strong submartingale satisfying one of the two properties stated in Corollary 4.3.7 (ii). Then for any $A \in \mathcal{A}$,

$$V_A = \lim_{m \to \infty} \lim_{n \to \infty, n \geq m} V^n_{g_m(A)}$$

in the weak L^1 topology, where V is the *-predictable *-compensator of X. V is adapted if Assumption 4.4.3 and the conditions of Lemma 1.4.3 hold.

Proof. As mentioned previously, we shall prove only (ii):

For each $t \in T$ and $F \in \mathcal{F}$ let

$$Y^n(F, t) = E[I_F \mid \mathcal{G}^*_{\mathcal{C}^n_t}] = \sum_{C \in \mathcal{C}^\ell(\mathcal{A}_n)} E[I_F \mid \mathcal{G}^*_C] I_C(t).$$

Since $(\mathcal{C}^n_t)$ is decreasing in n, $(\mathcal{G}^*_{\mathcal{C}^n_t})$ is increasing and $\lim_n Y^n(F, t) = Y(F, t) = E[I_F \mid H^*_t]$ almost surely. By Assumption 4.4.2, $Y(F, t)$ is well defined and $\mathcal{P}^*$-measurable and we may assume that $Y(F, t) \geq 0$ for all $t \in T$ and $\omega \in \Omega$.

We recall that μ_X can be extended to a σ-additive measure on $\mathcal{P}^*$. Therefore, for any $m \in \mathbf{N}$, $C \in \mathcal{C}_m$, $F \in \mathcal{F}$ we may define

$$\sigma_C(F) = \int_{\Omega \times C} Y(F, t) d\mu_X.$$

By dominated convergence, if $A \in \mathcal{A}_m$

$$\begin{aligned}
\sigma_A(F) &= \lim_{n \to \infty} \int_{\Omega \times A} Y^n(F, t) d\mu_X. \\
&= \lim_{n \to \infty, n \geq m} \sum_{C \in \mathcal{C}^\ell(\mathcal{A}_n), C \subseteq A} E[E[I_F \mid \mathcal{G}^*_C] X_C] \\
&= \lim_{n \to \infty, n \geq m} \sum_{C \in \mathcal{C}^\ell(\mathcal{A}_n), C \subseteq A} E[E[X_C \mid \mathcal{G}^*_C] I_F] \\
&= \lim_{n \to \infty, n \geq m} I_F V^n_A.
\end{aligned}$$

Then by the theorem of Hahn-Vitali-Saks, there exists a $\mathcal{F}$-measurable integrable random variable $\overline{V}_A$ such that $\sigma_A(F) = E[I_F \overline{V}_A]$ for each $F \in \mathcal{F}$, and $(V^n_A)_{n \geq m}$ converges to $\overline{V}_A$ in the weak L^1 topology. $\overline{V}_A$ will be $\mathcal{F}_A$-measurable if V^n_A is for each $n \geq m$. $\overline{V}$ can be (uniquely) extended to $\mathcal{C}_m$ and it is easily seen that $E[I_F \overline{V}_C] = \int_{\Omega \times C} Y(F, t) d\mu_X \geq 0$, so $\overline{V}_C \geq 0$ a.s. for every $C \in \mathcal{C}_m$ and for all $m \in \mathbf{N}$. Thus, there exists an event $\Omega_0 \subseteq \Omega$

such that $P(\Omega_0) = 1$ and $\overline{V}_C(\omega) \geq 0$ for every $C \in \cup_m \mathcal{C}_m$ and $\omega \in \Omega_0$. Now define for arbitrary $A \in \mathcal{A}$,

$$V_A = \lim_{m \to \infty} \overline{V}_{g_m(A)} I_{\Omega_0}.$$

Just as in the proof of Theorem 4.3.6, it is straightforward to show that V is an increasing process which is adapted if the conditions of Lemma 1.4.3 hold, by outer-continuity of the filtration. Clearly, V_A is the weak L^1 limit in the statement of the theorem and $V_A = \overline{V}_A$ a.s. if $A \in \cup_m \mathcal{A}_m$.

We now show that $\mu_X = \mu_V$ on $\mathcal{P}^*$. Let $C = A \setminus B \in \mathcal{C}$, $A \in \mathcal{A}$, $B \in \mathcal{A}(u)$ be a maximal representation of C, and define $C_m = g_m(A) \setminus g_m(B)$. Then $\mathcal{G}_C^* \subseteq \mathcal{G}_{C_m}^*$ since if $D \cap C = \emptyset$, $D \in \mathcal{A}(u)$, then $D \subseteq B \subseteq g_m(B)$ and so $D \cap C_m = \emptyset$. Now, by L^1-outer continuity and additivity of X and V, if $F \in \mathcal{G}_C^*$,

$$\begin{aligned}
E[I_F X_C] &= \lim_{m \to \infty} E[I_F X_{C_m}] \\
&= \lim_{m \to \infty} \lim_{n \to \infty, n \geq m} E[I_F X_{C_m}^n] \\
&= \lim_{m \to \infty} \lim_{n \to \infty, n \geq m} E[I_F V_{C_m}^n] \text{ by Lemma 4.4.5 (ii)} \\
&= E[I_F V_C].
\end{aligned}$$

Thus, $\mu_X = \mu_V$ on $\mathcal{P}^*$.

Finally, we show that V is *-predictable. Let $F \in \mathcal{F}$ and $C \in \mathcal{C}$ and define C_m as above.

$$\begin{aligned}
E[I_F V_C] &= \lim_m E[I_F V_{C_m}] \\
&= \lim_m \lim_{n \geq m} E[I_F V_{C_m}^n] \\
&= \lim_m \sigma_{C_m}(F) \\
&= \lim_m \int_{\Omega \times C_m} Y(F,t) d\mu_X \\
&= \lim_m \int_{\Omega \times C_m} Y(F,t) d\mu_V \\
&= \lim_m \int_{\Omega \times C_m} E[I_F \mid \mathcal{H}_t^*] d\mu_V \\
&= \int_{\Omega \times T} E[I_F \mid \mathcal{H}_t^*] I_C d\mu_V,
\end{aligned}$$

where the last equality follows by dominated convergence. This completes the proof of *-predictability.

Thus, V is the *-predictable increasing process of the Doob-Meyer decomposition, and it is the weak L^1-limit in the statement of the theorem. □

Analogously, we have the following:

Theorem 4.4.8 *If Assumptions 4.3.1 and 4.3.3 hold and if X is an L^2-monotone outer-continuous strong martingale such that the class*

$$\{ \sum_{C \in \mathcal{C}^\ell(\mathcal{A}')} E[(X_C)^2 \mid \mathcal{G}_C^*] : \mathcal{A}' \text{ a finite sub-semilattice of } \mathcal{A}\}$$

is uniformly integrable, then for any $A \in \mathcal{A}$,

$$Q_A = \lim_{m \to \infty} \lim_{n \to \infty, n \geq m} Q_{g_m(A)}^n$$

*in the weak L^1 topology, where Q is the *-predictable *-quadratic variation process of X. Q is adapted if Assumption 4.4.3 and the conditions of Lemma 1.4.3 hold.*

4.5 Point Processes and Compensators

The simplest example of a process which is both a strong and a weak submartingale with respect to any filtration is a *point process*.

Definition 4.5.1 *A process $N = \{N_A; A \in \mathcal{A}\}$ is a point process if it is an increasing process taking its values in $\mathbf{N}$, and almost surely for any $t \in T$, $N_{\{t\}} \leq 1$.*

Comment: This last requirement makes sense since N is increasing and therefore has a unique extension as a measure on the Borel sets of T. Sometimes, a point process satisfying $N_{\{t\}} \leq 1$ is referred to as a simple point process. As we consider only point processes with this property, the word 'simple' will not be used. We will introduce a more stringent condition known as 'strict simplicity' when we deal with the weak convergence of sequences of point processes, but this concept is not required here.

Regarding a point process as both a weak and a strong submartingale, we can calculate both compensators and *-compensators and we consider some important examples. In what follows, recall that $\mathcal{F}^0$ and $\mathcal{F}^r$ denote the minimal filtration and minimal outer-continuous filtration, respectively.

The Poisson and Cox Processes

In the case of the Poisson and Cox processes, both compensators are the same. The following theorem is essentially Propositions 3.4.6 and 3.4.8.

Theorem 4.5.2 *(i) Let N be a Poisson process with mean measure Λ. Then N is both a submartingale and a strong submartingale. Λ is a compensator and a *-compensator for N with respect to both $\mathcal{F}^0$ and $\mathcal{F}^r$, and is adapted, predictable and *-predictable. Λ is a quadratic variation and a *-quadratic variation for the (strong) martingale $N - \Lambda$.*
(ii) The same statements are true if N is a Cox process driven by the $\mathcal{F}_\emptyset$-measurable measure Λ.

Although for the Poisson process both compensators are the same, clearly this is not true in general. This raises the question of which type of compensator is most useful. In the classical theory of point processes on $\mathbf{R}_+$, it is well known that under certain conditions on the probability space and filtration, the compensator of a simple point process determines its distribution (Jacod (1975)). As we shall see, this is not true for point processes on $\mathbf{R}_+^2$, as many distinct point processes may have the same compensator with respect to the minimal filtration. We shall let $T = [0,1]^2$ and $\mathcal{A} = \{[0,s] \times [0,t] : 0 \leq s, t \leq 1\}$. Then the Poisson point process with mean measure equal to Lebesgue measure has compensator $\Lambda_{[0,s] \times [0,t]} = st$. However, Λ is also the compensator of the point processes in the following counterexamples:

Counterexample 1: Let $L_0, L_1, L_2, ...$ be i.i.d. unit rate Poisson processes on $[0,1]$. Denote the time of the kth jump of L_i by T_k^i, $k = 1, 2, ...$ Now, let the locations of the points of N be $\{(T_i^0, T_k^i); i, k \geq 1\}$ (i.e., $N_{[0,s] \times [0,t]} = \sum_{i=1}^\infty \sum_{k=1}^\infty I_{T_i^0 \leq s} I_{T_k^i \leq t}$). (See Figure 4.1.)

Counterexample 2: All the points of N lie on diagonal lines connecting $(U_i, 0)$ and $(0, U_i)$ or connecting $(V_j, 1)$ with $(1, V_j)$, where $(U_1, U_2, ...)$ and $(V_1, V_2, ...)$ are independent Poisson processes on $[0,1]$ with rate $2^{-1/2}$. The points on each diagonal line then form independent unit rate Poisson processes (see Figure 4.2).

Thus, it is clear that the compensator with respect to the minimal filtration is of limited use. It is shown in Ivanoff and Merzbach (1990a) that given one particular filtration which is strictly larger than $\mathcal{F}^0$, there is a complete characterization of compensators of point processes on $\mathbf{R}_+^2$. However, the definition of this filtration

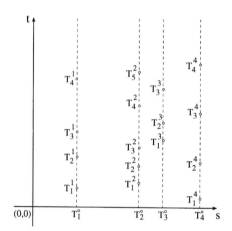

Figure 4.1

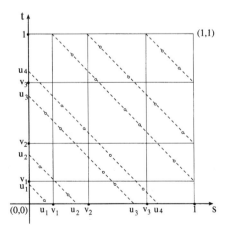

Figure 4.2

relies on the axes and so cannot be extended to our more general framework.

Finally, we consider the *-compensator. The question of whether the *-compensator with respect to the minimal filtration determines the distribution of a point process is still open even in the case of $\mathbf{R}_+^2$. However, it will be proven in the next chapter that the answer is affirmative for any set-indexed Poisson process. As well, we shall see in Part 2 that the behaviour of the *-compensators of a sequence of set-indexed point processes can provide sufficient conditions for weak convergence of the sequence. Thus, henceforth we shall restrict our attention to the *-compensator.

The Single Jump Process

A simple example (discussed in Ivanoff and Merzbach (1997)) which plays an important rôle in the theory of empirical processes is the single jump process; this is a process which has a single jump at some (random) point $Y \in T$. We then have that $N_A = I_{\{Y \in A\}}$. For this example, we shall assume that T is a compact subset of $\mathbf{R}^d$, say $T = [-K, K]^d$, for $K < \infty$. The family $\mathcal{A}$ can be chosen to be either the class of rectangles $\{[0, \mathbf{x}], \mathbf{x} \in T\}$ or the class of lower sets. For any point $t = (t_1, ..., t_d) \in T$, let $R_t = \prod_{i=1}^d \Delta t_i$, where
$$\Delta t_i = \begin{cases} [0, t_i] & \text{if } t_i > 0 \\ \{0\} & \text{if } t_i = 0 \\ [t_i, 0] & \text{if } t_i < 0 \end{cases}$$
R_t may be seen as the 'strict past' of t and it is the smallest closed rectangle containing both the origin and the point t. In the same way, we may define the 'strict future' of t: $S_t = T \cap \prod_{i=1}^d \Delta' t_i$, where
$$\Delta' t_i = \begin{cases} \mathbf{R}_+ \setminus \Delta t_i & \text{if } t_i > 0 \\ \mathbf{R}_- \setminus \Delta t_i & \text{if } t_i < 0. \end{cases}$$
S_t is not defined if $t_i = 0$ some $i, 1 \leq i \leq d$.

Now, let Y be a T-valued random variable and let F be its distribution function: $F(t) = P(Y \leq t), t \in T$. F will also be viewed as a measure, and so for any Borel set $F(A) = \int_A dF$. Recall that $N_A = I_{\{Y \in A\}}$ and note that the minimal filtration generated by N is monotone outer-continuous (this is analogous to $\mathbf{R}_+$). A *-compensator of N is given in the following theorem. In the case in which T is the two-dimensional quadrant $\mathbf{R}_+^2$ and $\mathcal{A} = \{R_t : t \in \mathbf{R}_+^2\}$, a similar result was obtained by Al-Hussaini and Elliott (1984).

POINT PROCESSES AND COMPENSATORS

Theorem 4.5.3 *Let N be the single jump process as defined above. Then if $\mathcal{F}$ is the minimal filtration, V is a *-compensator of N, where*

$$V_A = \int_{A \cap R_Y} (F(\overline{S_u}))^{-1} dF(u). \tag{4.4}$$

(By convention, we set $(F(\overline{S_u}))^{-1} = 0$ if $F(\overline{S_u}) = 0$.)

Proof. Suppose first that $C = \prod_{i=1}^{d} C_i$, where C_i is of either the form $(a_i, b_i]$, $0 \leq a_i < b_i$, or $[b_i, a_i)$, $b_i < a_i \leq 0$. It is easily seen that $E[N_c \mid \mathcal{G}_C^*] = I_{\{Y \in S_a\}} F_C(F(S_a))^{-1}$. Indeed, $\cup_{t \notin S_a} \mathcal{F}_{R_t} \subseteq \mathcal{G}_C^*$ and $\{Y \in S_a\} = \cap_{t \notin S_a} \{Y \notin R_t\}$. Since this last set can be expressed as a countable intersection, we obtain that $\{Y \in S_a\} \in \mathcal{G}_C^*$.

Now,

$$E[V_C \mid \mathcal{G}_C^*] = \frac{I_{\{Y \in S_a\}}}{F(S_a)} \int_{S_a} \int_{C \cap R_v} (F(\overline{S_u}))^{-1} dF(u) dF(v).$$

By changing the order of integration, this double integral is equal to

$$\int_C (F(\overline{S_u}))^{-1} \int_{\overline{S_u}} dF(v) dF(u) = F(C).$$

Therefore, $E(N_C - V_C \mid \mathcal{G}_C^*) = 0$. For $C = \prod_{i=1}^{d} C_i$ where $C_i = [0, b_i]$ or $[b_i, 0]$ for one or more i, the result follows since both N and V put no mass on $\{t \in T : t_i = 0, \text{ some } i, 1 \leq i \leq d\}$.

The next step is to consider sets C which are finite disjoint unions $C = \cup_{i=1}^{k} C_i$ of sets of the above type. Then, by additivity and the fact that $\mathcal{G}_C^*$ decreases as C increases,

$$E[N_C - V_C \mid \mathcal{G}_C^*] = \sum_{i=1}^{k} E[E(N_{C_i} - V_{C_i} \mid \mathcal{G}_{C_i}^*) \mid \mathcal{G}_C^*] = 0.$$

Finally, we suppose that $C = A \backslash \cup_{i=1}^{k} B_i$ is a maximal representation (in T) of an arbitrary set in $\mathcal{C}$. Let $C_m = g_m(A) \backslash \cup_{i=1}^{k} g_m(B_i)$. Since N and V may be regarded as measures, $N_{C_m} - V_{C_m} \to N_C - V_C$ almost surely and in L^1. By uniform integrability, it follows that $E[N_C - V_C \mid \mathcal{G}_C^*] = \lim_m E[N_{C_m} - V_{C_m}] \mid \mathcal{G}_C^*]$. Since we have a maximal representation of C, if $D \in \mathcal{A}$, $D \cap C = \emptyset$, then $D \subseteq \cup_{i=1}^{k} B_i \subseteq \cup_{i=1}^{k} g_m(B_i)$ for each m, so $D \cap C_m = \emptyset$. Therefore, $\mathcal{G}_C^* \subseteq \mathcal{G}_{C_m}^*$, and so $E[N_{C_m} - V_{C_m} \mid \mathcal{G}_C^*] = E[E(N_{C_m} - V_{C_m} \mid \mathcal{G}_{C_m}^*) \mid \mathcal{G}_C^*] = 0$, completing the proof. $\square$

CHAPTER 5

Martingale Characterizations

The goal of this chapter is to extend to set-indexed processes the classic martingale characterizations of the Poisson process and the Brownian motion on $\mathbf{R}_+$ due, respectively, to Watanabe and Lévy.

The Watanabe characterization states that any simple point process on $\mathbf{R}_+$ which has a continuous and deterministic compensator is a Poisson process. This was generalized to $\mathbf{R}_+^2$ under the restriction that the point process have no more than one jump on each line parallel to one of the axes (cf. Ivanoff (1985) and Merzbach and Nualart (1986)). Similar characterizations for set-indexed processes were obtained without using martingale techniques by Papangelou (1974) and subsequently by Kallenberg (1978) and van der Hoeven (1983). However, these results required taking the conditional expectation of the point process at any given set with respect to a filtration which takes into consideration the entire complement of the set. The martingale characterization given here is more general, in that $\mathcal{A}$ can be much smaller than the class of sets required by Papangelou and Kallenberg. It also generalizes the previous martingale results for point processes on $\mathbf{R}_+^d$ in that the process can be concentrated on a lower-dimensional subspace.

The Lévy characterization states that any continuous martingale on $\mathbf{R}_+$ whose quadratic variation process is the deterministic function $v(s) = s$ is a Brownian motion. Extensions to strong martingales on $\mathbf{R}_+^2$ were proven first by Wong and subsequently by Zakai (1981). Nualart (1981a) proved a similar result for planar martingales with orthogonal increments, and showed that the result is no longer valid for martingales which are not strong martingales. The martingale characterization of set-indexed Brownian motion which will be proven here is in fact more general than that of Wong and Zakai, even in the two-parameter case.

The key ingredient for both characterizations is the use of a *flow*; that is, an increasing function from an interval $[a, b]$ to either $\mathcal{A}(u)$ or $\tilde{\mathcal{A}}(u)$. In each case, the judicious construction of a flow with particular properties permits us to reduce the general problem to one dimension. Flows will be discussed thoroughly in the next section. The martingale characterizations will be given in the two subsequent sections. The Brownian motion and the Poisson process were studied in Ivanoff and Merzbach (1996) and (1994), respectively.

As both the Brownian motion and Poisson processes are determined by their finite-dimensional distributions on compact subsets of T, without loss of generality *we shall assume for the remainder of this chapter that T is compact and that $T \in \mathcal{A}_n(u)$ $\forall n$*. As well, we assume that Assumption 1.1.7 holds: i.e., that the class of all left-neighbourhoods generated by the semi-lattices $\mathcal{A}_n, n \geq 1$, is a dissecting system for T.

5.1 Flows

As mentioned above, the problem of characterizing the Brownian motion and the Poisson process may be reduced to a one-dimensional problem. As is clear from Definitions 3.4.2 and 3.4.4, these two processes are characterized by their finite-dimensional distributions over disjoint sets in $\mathcal{C}$. This is exactly the definition of Brownian motion, and the class $\mathcal{C}$ suffices for the Poisson process since it is a semiring which generates the Borel sets of T (see, for example, Daley and Vere-Jones (1988)). Now, for any (additive) $\mathcal{A}$-indexed process X, the finite-dimensional distributions over $\mathcal{C}$ are completely determined by the joint distributions of the increments of X over the class $\{C_0 = \emptyset', C_1, ..., C_k\}$ of left-neighbourhoods of an arbitrary finite sub-semilattice $\mathcal{A}'$ of $\mathcal{A}$. Without loss of generality, we may assume that $\mathcal{A}' = \mathcal{A}_1$, and equip $\mathcal{A}_1 = \{\emptyset' = A_0, A_1, ..., A_k\} = \{\emptyset' = A_{1,0}, A_{1,1}, ..., A_{1,k_1}\}$ with a numbering consistent with the strong past, and for $i = 1, ..., k$, $C_i = A_i \setminus \cup_{j=0}^{k-1} A_j$. Recall that we always assume that $X_{\emptyset'} = 0$.

We will identify the finite dimensional distributions by associating to X a process Y defined on $[0, k]$ such that $Y_0 = X_{\emptyset'} = 0$, and $X_{C_i} = Y_i - Y_{i-1}$, $i = 1, ..., k$. This is done via a *flow*. The definition of a flow is as follows:

Definition 5.1.1 *Let $S = [a, b] \subseteq \mathbf{R}$. An increasing function $f : S \to \tilde{\mathcal{A}}(u)$ is called a flow. A flow f is right-continuous if*

$$f(s) = \cap_{v>s} f(v), \forall s \in [a, b),$$

and $f(b) = \overline{\cup_{u<b} f(u)}$. A flow f is continuous if it is right-continuous

and
$$f(s) = \overline{\cup_{u<s} f(u)} \ \forall s \in (a,b).$$

Given a set-indexed process X and a flow f, we wish to define a process Y indexed by $S = [a,b]$ as follows: $Y_s = X_{f(s)} \ \forall s \in S$. Y is well defined provided that $f : S \to \mathcal{A}(u)$. However, it is not true in general that an integrable process X defined on $\mathcal{A}(u)$ can be extended to an integrable process on $\tilde{\mathcal{A}}(u)$ using outer limits. As has already been observed, in the case of the Brownian motion on $[0,1] \times [0,1]$ with $\mathcal{A}$ the rectangles, the sample paths are almost surely unbounded on $\tilde{\mathcal{A}}(u)$, the class of lower sets. We do have the following:

Lemma 5.1.2 *Let X be a strong martingale and let f be a flow defined on $[0,b]$. Then the one-parameter process $\{Y_s = X_{f(s)} : s \in [0,b]\}$ is a martingale with respect to the filtration $\mathcal{H} = \{\mathcal{H}_s = \mathcal{F}_{f(s)}\}$ in either of the following situations:*
(i) $f : [0,b] \to \mathcal{A}(u)$, or
(ii) X is the difference between two integrable increasing processes and $\mathcal{F} = \mathcal{F}^r$.

Proof.
(i) Let $0 \le s < r \le b$ and suppose $f(s) = B, f(r) = B' \in \mathcal{A}(u)$. Let $B' = \cup_{i=1}^k A_i$, $A_i \in \mathcal{A}$, $i = 1, \ldots, k$. Without loss of generality, assume that for each $i, j \in \{1, \ldots, k\}$, there exists $\ell \in \{1, \ldots, k\}$ such that $A_i \cap A_j = A_\ell$. (This representation is not extremal.) Then we may write $B' \setminus B = \cup_{i=1}^k A_i \setminus B = \cup_{i=1}^k \left(A_i \setminus \cup_{j : A_i \not\subseteq A_j} A_j \cup B\right) = \cup_{i=1}^k C_i$, where $C_1, \ldots, C_k$ are disjoint sets in $\mathcal{C}$. Clearly, $\mathcal{F}_B \subseteq \mathcal{G}_{C_i}^*$, $i = 1, \ldots, k$. Thus,

$$\begin{aligned} E(Y_r - Y_s | \mathcal{H}_s) &= E(X_{B'} - X_B | \mathcal{F}_B) \\ &= E\left(\sum_{i=1}^k X_{C_i} \Big| \mathcal{F}_B\right) \\ &= E\left(\sum_{i=1}^k E\left[X_{C_i} | \mathcal{G}_{C_i}^*\right] \Big| \mathcal{F}_B\right) \\ &= 0, \end{aligned}$$

and Y is a martingale.

(ii) Now suppose that $f(s) = B, f(r) = B' \in \tilde{\mathcal{A}}(u)$. Since an increasing process can be extended to a measure on the σ-algebra generated by $\mathcal{A}$, X_B and $X_{B'}$ are well-defined and integrable.

Therefore,

$$
\begin{aligned}
E(Y_r - Y_s|\mathcal{H}_s) &= E(X_{B'} - X_B|\mathcal{F}_B) \\
&= \lim_n E\left(X_{g_n(B')} - X_{g_n(B)}|\ \mathcal{F}_B\right)
\end{aligned}
$$

(by outer – continuity and uniform integrability of increasing processes)

$$
= \lim_n \lim_{m \geq n} E\left(X_{g_n(B')} - X_{g_n(B)}|\ \mathcal{F}_{g_m(B)}\right)
$$

(by the reverse martingale convergence theorem)

$$
= 0 \quad \text{(by (i) and since } g_m(B) \subseteq g_n(B) \text{ if } m \geq n\text{)}.
$$

□

Comments 5.1.3 1. It is a simple consequence of Lemma 1.4.1 that the filtration $\mathcal{H} = \{\mathcal{H}_s\} = \{\mathcal{F}_{f(s)}\}$ is right-continuous if $\mathcal{F} = \mathcal{F}^r$ and if f is a right-continuous flow.

2. Part (ii) of the preceding theorem can be stated as follows: If X is a strong martingale which is the difference between two integrable increasing processes and if $\mathcal{F} = \mathcal{F}^r$, then for any sets $B \subseteq B'$, $B, B' \in \check{\mathcal{A}}(u)$,

$$
E(X_{B'} - X_B \mid \mathcal{F}_B) = 0.
$$

We now begin to study the construction of flows with particular properties necessary to identify the Brownian motion and the Poisson process. In both cases, given a particular increasing deterministic function Λ defined on $\mathcal{A}$ and the class $\{C_0, C_1, ..., C_k\}$ of left-neighbourhoods of the finite sub-semilattice $\mathcal{A}_1$ of $\mathcal{A}$, the goal is to construct a flow f on $[0, k]$ such that $f(0) = \emptyset'$, $C_i = f(i) - f(i-1)$, $i = 1, ..., k$ and the generated function $\lambda = \Lambda \circ f$ on $[0, k]$ is continuous. However, in light of the comments preceding Lemma 5.1.2, in the case of Brownian motion we must have that $f : [0, k] \to \mathcal{A}(u)$ and we will also require that $F \circ f$ be continuous for any monotone inner- and outer-continuous $\mathcal{A}$-indexed function F (inner-continuity is defined below). These restrictions will not be necessary in the case of the Poisson process, and so we will give two different constructions.

First Construction (Ivanoff and Merzbach (1996)):

We begin by defining monotone inner-continuity:

Definition 5.1.4 *A function* $F : \mathcal{A} \to \mathbf{R}$ *is monotone inner-*

FLOWS   101

continuous on $\mathcal{A}$ if for any increasing sequence (A_n) of sets in $\mathcal{A}$ such that $\overline{\cup_n A_n} = A \in \mathcal{A}$, then $\lim_n F(A_n) = F(A)$.

Remark 5.1.5 We note that inner- and outer-continuity on $\mathcal{A}$ are defined in terms of increasing and decreasing sequences of sets in $\mathcal{A}$. Thus, if $f : [a, b] \to \mathcal{A}$ is a continuous flow and if $F : \mathcal{A} \to \mathbf{R}$ is monotone inner- and outer-continuous on $\mathcal{A}$, then obviously $F \circ f$ is continuous.

It is not clear that Remark 5.1.5 above remains true if the range of the flow f is $\mathcal{A}(u)$, so we will now proceed to show that such a flow may be constructed, which also has the other properties described above.

The first step is the following lemma:

Lemma 5.1.6 *Let $A, B \in \mathcal{A}_1, A \subset B$, and B minimal (in $\mathcal{A}_1$) with respect to A. Then there exists a continuous flow $f : [0, 1] \to \mathcal{A}$ with $f(0) = A$ and $f(1) = B$.*

Proof. We begin by defining $f(0) = A, f(1) = B$. Continuing inductively, suppose there exist points $s = i/r, v = (i+1)/r, i, r \in \mathbf{Z}_+, r > 0$, such that $f(s) = S$, $f(v) = V$, $S, V \in \mathcal{A}_{n-1}$, and V is minimal (in $\mathcal{A}_{n-1}$) with respect to S. Then, using Lemma 1.1.9, it is possible to define a chain $\{S = A_0 \subset A_1 \subset \ldots \subset A_k = V\}$ in $\mathcal{A}_n$ such that A_i is minimal in $\mathcal{A}_n$ with respect to A_{i-1}. (Clearly $k = k(S, V, \mathcal{A}_n)$, but the dependence is supressed for ease of notation.) Now define $f((i/r) + (j/rk)) = A_j, j = 0, \ldots, k$. We continue in this way, for every n.

Let $\mathcal{I}$ be the set of points in $[0, 1]$ on which f has been defined. We claim that $\mathcal{I}$ is dense in $[0, 1]$. That this is a result of the Definition 1.1.1 of an indexing collection may be seen as follows: If $C \subset D, C, D \in \mathcal{A}_n$ and D is minimal (in $\mathcal{A}_n$) with respect to C, then there exists $m > n$ and $E \in \mathcal{A}_m$ such that $C \subset E \subset D$. This follows, since by Definition 1.1.1 (d) and (e), $C = \cap_m g_m(C)$ so $\exists m$ sufficiently large that $C \subset g_m(C) \cap D \subset D$. If $\mathcal{I}$ is not dense in $[0, 1]$, there exists a closed interval $[a, b]$ in $[0, 1]$ such that $\mathcal{I} \cap (a, b) = \emptyset$ and $a = \sup_{s \in \mathcal{I}, s \leq a} s, b = \inf_{v \in \mathcal{I}, v \geq b} v$. Clearly, we may choose n and $C, D \in \mathcal{A}_n$ and D minimal (in $\mathcal{A}_n$) with respect to C such that $C = f(s), D = f(v), s \leq a < b \leq v, a - s < (b-a)/2, v - b < (b-a)/2$. But by the preceding discussion, there exists m such that $\mathcal{A}_m$ contains at least one set strictly between C and D. In this case, there exists a set E in $\mathcal{A}_m$ and a real number $e \in (a, b)$ such that $f(e) = E$. Thus, by contradiction, $\mathcal{I}$ is dense in $[0, 1]$.

We will now show that for every $r \in (0,1)$,
$$L = \overline{\cup_{s \in \mathcal{I}, s < r} f(s)} = \cap_{s \in \mathcal{I}, s > r} f(s) = U.$$
Suppose that $L \subset U$ and $r \in \mathcal{I}^c$. Then $\exists n$ such that $U \not\subseteq g_n(L)$. Fix such a number n. Let $u = \max\{s < r : f(s) \in \mathcal{A}_n\}, v = \min\{s > r : f(s) \in \mathcal{A}_n\}$. Since $\mathcal{A}_n$ is finite and $r \in \mathcal{I}^c$, we have $u < v$ and $f(v)$ is minimal (in $\mathcal{A}_n$) with respect to $f(u)$. But $g_n(L) \cap f(v) \in \mathcal{A}_n$, $f(v) \not\subseteq g_n(L)$ (since $U \subseteq f(v)$), and so
$$f(u) \subseteq L \subset g_n(L) \cap f(v) \subset f(v).$$
But this contradicts the fact that $f(v)$ is minimal (in $\mathcal{A}_n$) with respect to $f(u)$. Thus, $L = U$. A similar argument may be used to prove that $L = U$ if $r \in \mathcal{I}$. As well, it is easily seen that $A = f(0) = \cap_{s \in \mathcal{I}, s > 0} f(s)$ and that $B = f(1) = \overline{\cup_{s \in \mathcal{I}, s < 1} f(s)}$.

Finally, we may define for all $r \in [0,1]$,
$$f(r) = \cap_{s \in \mathcal{I}, s > r} f(s).$$
Clearly f is a continuous flow satisfying the statement of the lemma. □

We now complete the construction:

Lemma 5.1.7 *Let $\mathcal{A}_1 = \{\emptyset' = A_0, ..., A_k\}$ be any finite sub-semilattice of $\mathcal{A}$ equipped with a numbering consistent with the strong past. Then there exists a continuous flow $f : [0,k] \to \mathcal{A}(u)$ such that the following are satisfied:*

1. $f(i) = \cup_{j=0}^{i} A_j$ for $i = 0, ..., k$.
2. *Each left-neighbourhood C generated by $\mathcal{A}_1$ is of the form $C = f(i) \setminus f(i-1), 1 \leq i \leq k$.*
3. *If $F : \mathcal{A}(u) \to \mathbf{R}$ is additive and monotone inner- and outer-continuous on $\mathcal{A}$, then $F \circ f : [0, k] \to \mathbf{R}$ is continuous.*

Proof. For $1 \leq i \leq k$, let $f_i : [0,1] \to \mathcal{A}$ be a continuous flow with $f(0) = A_{j_i}$ and $f(1) = A_i$, where A_{j_i} is as defined in Lemma 1.1.9 (i.e., $j_i < i$, and A_i is minimal with respect to A_{j_i}). Obviously, $A_{j_1} = \emptyset' = A_0$. Now let $f(s) = f_1(s)$, for $s \in [0,1]$. For $i-1 \leq s \leq i$, let
$$f(s) = f(i-1) \cup f_i(s - i + 1).$$
It is easily seen that $f(i) = \cup_{j=0}^{i} A_j$ (proving 1), that $f(s) \in \mathcal{A}(u), \forall s, 0 \leq s \leq k$, and that f is continuous on $\mathcal{A}(u)$. All of the left-neighbourhoods of sets in $\mathcal{A}_1$ are of the form $\cup_{j=0}^{i} A_j \setminus \cup_{j=0}^{i-1} A_j = f(i) \setminus f(i-1), 1 \leq i \leq k$, proving 2.

To prove 3, we observe that by construction if $s_n \downarrow s$, then for all n sufficiently large, $f(s_n) \setminus f(s) = f_{[s]+1}(s_n - [s]) \setminus f_{[s]+1}(s - [s])$ ($[s]$ is the integer part of s). Since $f_i : [0,1] \to \mathcal{A}$, it follows from Remark 5.1.5 and additivity of F that $F \circ f$ is right-continuous. Left-continuity follows in a similar manner. □

Second Construction (Ivanoff and Merzbach (1994)):

In the case of the Poisson process, since any increasing process defines a (random) measure on $\mathcal{B}$, the process is well-defined on $\tilde{\mathcal{A}}(u)$. This permits us to define a more general type of flow f such that $\Lambda \circ f$ will be continuous even if Λ is no longer inner-continuous.

The first step in this construction is to iteratively number the sublattices $\mathcal{A}_n$ in a manner consistent with the strong past as follows:

$\mathcal{A}_1 = \{\emptyset' = A_{1,0}, \ldots, A_{1,k_1}\}$. Now continue inductively as follows: given a numbering on $\mathcal{A}_r$ consistent with the strong past, say $\mathcal{A}_r = \{A_{r,0}, \ldots, A_{r,k_r}\}$, $\mathcal{A}_{r+1}$ is partitioned as defined below:

$$\mathcal{A}_{r+1}^0 = \{\emptyset'\}$$
$$\mathcal{A}_{r+1}^1 = \{A \in \mathcal{A}_{r+1} : A \subseteq A_{r,1}, A \neq \emptyset'\}$$
$$\mathcal{A}_{r+1}^i = \{A \in \mathcal{A}_{r+1} : A \subseteq A_{r,i}, A \not\subseteq A_{r,j}, j = 0, \ldots, i-1\},$$
$$i = 2, \ldots, k_r$$
$$\mathcal{A}_{r+1}^{k_r+1} = \{A \in \mathcal{A}_{r+1} : A \not\subseteq A_{r,j}, j = 0, \ldots, k_r\}.$$

Assign to each class $\mathcal{A}_{r+1}^i$ a numbering consistent with the strong past, say $\{A_{r+1,1}^i, \ldots, A_{r+1,\ell_i}^i\}$, $i = 1, \ldots, k_r + 1$. Finally, give to $\mathcal{A}_{r+1}$ the numbering $\mathcal{A}_{r+1} = \{A_{r+1,0}, \ldots, A_{r+1,k_{r+1}}\}$ where $A_{r+1,0} = \emptyset'$ and for $j \geq 1$,

$$A_{r+1,j} = A_{r+1,h}^i \quad \ell_1 + \ldots + \ell_{i-1} < j \leq \ell_1 + \ldots + \ell_i,$$
$$h = j - (\ell_1 + \ldots + \ell_{i-1})$$
$$j = 1, \ldots, k_{r+1} \ (k_{r+1} = \ell_1 + \ldots + \ell_{k_r+1})$$
(If $i = 1$ set $\ell_1 + \cdots + \ell_{i-1} = 0$.)

It is easy to verify that the numbering

$$\mathcal{A}_{r+1} = \{A_{r+1,0}, \ldots, A_{r+1,k_{r+1}}\}$$

is consistent with the strong past.

Note that heuristically speaking, the above numbering is carried out in such a manner that when the sets in $\mathcal{A}_{r+1}$ are numbered, all those sets contained in $A_{r,i}$ are numbered before numbering those sets which are contained in $A_{r,i+1}$, but not contained in $A_{r,i}$. Also, if $A \in \mathcal{A}_{r+1}^i$, then the left-neighbourhood (in $\mathcal{A}_{r+1}$) of A is contained in the left-neighbourhood (in $\mathcal{A}_r$) of $A_{r,i}$. This point is used in the proof of the following lemma which is extremely important for the construction.

Lemma 5.1.8 Let $V = \{V_A, A \in \mathcal{A}\}$ be a monotone outer-continuous increasing (deterministic) function such that $V_{\{t\}} = 0 \ \forall \ t \in T$. Suppose that $A, B \in \tilde{\mathcal{A}}(u)$ and $A \subseteq B$. Then for every $v \in \mathbf{R}$ such that $V_A \leq v \leq V_B$, there exists a closed set $A_v \in \tilde{\mathcal{A}}(u)$ such that $A \subseteq A_v \subseteq B$ and $V_{A_v} = v$.

Proof. Suppose first that $A, B \in \mathcal{A}_m(u)$, for some m. Assume $V_A < v < V_B$. For $n \geq m$, number the sets in $\mathcal{A}_n$ iteratively as described above, omitting any sets in $\mathcal{A}_n$ which are either contained in A or not contained in B. Then we have a numbering $\{A_{n,1}, \ldots, A_{n,\ell_n}\}$ consistent with the strong past of all sets in $\mathcal{A}_n$ contained in B and not contained in A.
Let:

$$i(v,n) = \inf\{i : V_{A \cup \cup_{j=1}^i A_{n,j}} \geq v\} \text{ and}$$

$$A_{n,v} = A \cup \cup_{j=1}^{i(v,n)} A_{n,j} = A \cup \cup_{j=1}^{i(v,n)} C_{n,j},$$

where $C_{n,j}$ is the left-neighbourhood of $A_{n,j}$ in $\mathcal{A}_n$.

Note that this last representation is a union of disjoint sets. Thus

$$V_{A_{n,v}} = V_A + \sum_{j=1}^{i(v,n)} V_{C_{n,j}}.$$

Clearly, by the iterative nature of the numbering, $A_{n+1,v} \subseteq A_{n,v}$ and $C_{n+1,v} \subseteq C_{n,v}$, where $C_{n,v} = C_{n,i(v,n)}$. Let $A_v = \cap_n A_{n,v}$. A_v is closed and $A_v \in \tilde{\mathcal{A}}(u)$. By construction, $V_{A_v} \geq v$. If $V_{A_v} = v+\varepsilon$, for some $\varepsilon > 0$, then $V_{C_{n,v}} \geq \varepsilon \ \forall \ n$. But as noted above, $\{C_{n,v}\}_n$ is a nested sequence, and since $\{\mathcal{C}_n\}$ is a dissecting system, $C = \cap_n C_{n,v}$ contains at most one point, which implies that $V_C = 0$. Thus by contradiction, $V_{A_v} = v$.

For general $B \in \tilde{\mathcal{A}}(u)$, let $\mathcal{A}_B = \{A' \cap B : A' \in \mathcal{A}\}$, $\mathcal{A}_{Bn} = \{A' \cap B : A' \in \mathcal{A}_n\}$, and define $\mathcal{A}_B(u), \mathcal{A}_{Bn}(u), \mathcal{C}_B, \mathcal{C}_{Bn}$ and $\tilde{\mathcal{A}}_B(u)$ as before, using $\mathcal{A}_B$ and $\mathcal{A}_{Bn}$. Note that all sets in $\mathcal{A}_B, \mathcal{A}_B(u) \tilde{\mathcal{A}}_B(u)$ are in $\tilde{\mathcal{A}}(u)$. For any set $D \in \mathcal{A}_B$, define $g_{n,B}(D) = g_n(D) \cap B$, and extend $g_{n,B}$ as before to $\mathcal{A}_B(u)$ and $\tilde{\mathcal{A}}_B(u)$. Then $g_{n,B}$ satisfies the same properties as g_n, in the relative topology on B, and $\{\mathcal{C}_{Bn}\}$

is a dissecting system for B. If $A \subseteq B$, $A \in \tilde{\mathcal{A}}(u)$ and $V_A < v$, then there exists m sufficiently large that $V_{g_{m,B}(A)} < v$, and so the same proof as above shows that there exists $A_v \in \tilde{\mathcal{A}}(u)$ such that $A \subseteq g_{m,B}(A) \subseteq A_v \subseteq B$ and $V_{A_v} = v$.

This completes the proof. $\square$

We now are in a position to prove the following:

Lemma 5.1.9 *Let* $\mathcal{A}_1 = \{\emptyset' = A_0, ..., A_k\}$ *be any finite subsemilattice of $\mathcal{A}$ equipped with a numbering consistent with the strong past and let V be a deterministic monotone outer-continuous increasing function* $V : \mathcal{A} \to \mathbf{R}$ *such that* $V_{\{t\}} = 0 \; \forall t \in T$. *Then there exists a right continuous flow* $f : [0, k] \to \tilde{\mathcal{A}}(u)$ *such that the following are satisfied:*

1. $f(i) = \cup_{j=0}^{i} A_j$ *for* $i = 0, ..., k$.
2. *Each left-neighbourhood C generated by $\mathcal{A}_1$ is of the form* $C = f(i) \setminus f(i-1), 1 \le i \le k$.
3. $V \circ f : [0, k] \to \mathbf{R}$ *is continuous.*

Proof. For $i = 0, \ldots, k_1$, define $D_i = \cup_{j=0}^{i} A_{1,j} = \cup_{j=0}^{i} C_{1,j}$ where $C_{1,j}$ is the left-neighbourhood of $A_{1,j}$ in $\mathcal{A}_1$. Fix $i, 1 \le i \le k_1$. We shall define a right continuous flow $f_i : [0, 1] \to \tilde{\mathcal{A}}(u)$ with $f(0) = D_{i-1}$ and $f(1) = D_i$. Let $\tilde{f}_i(1/2)$ be any set in $\tilde{\mathcal{A}}(u)$ such that $D_{i-1} \subseteq \tilde{f}_i(1/2) \subseteq D_i$ and $V_{\tilde{f}_i(1/2)} = (V_{D_{i-1}} + V_{D_i})/2$ (the existence of such a set is ensured by the preceding lemma). Inductively, for $n > 1$, and $1 \le k \le 2^n$, $\tilde{f}_i(k \cdot 2^{-n}) = \tilde{f}_i(j \cdot 2^{-(n-1)})$ if $k = 2j$. If $k = 2j+1$, let $\tilde{f}_i(k \cdot 2^{-n})$ be any set in $\tilde{\mathcal{A}}(u)$ such that $\tilde{f}_i(j \cdot 2^{-(n-1)}) \subseteq \tilde{f}_i(k \cdot 2^{-n}) \subseteq \tilde{f}_i((j+1) \cdot 2^{-(n-1)})$ and $V_{\tilde{f}_i(k \cdot 2^{-n})} = (V_{\tilde{f}_i(j \cdot 2^{-(n-1)})} + V_{\tilde{f}_i((j+1) \cdot 2^{-(n-1)})})/2$. We now have that $\tilde{f}_i$ is an increasing function from the dyadics to a class of closed sets in $\tilde{\mathcal{A}}(u)$, which contain D_{i-1} and are contained in D_i. For any dyadic d, $V_{\tilde{f}_i(d)} = V_{D_{i-1}} + d(V_{D_i} - V_{D_{i-1}})$. Now, for arbitrary $u \in [0, 1]$, let $f_i(u) = \cap_{d > u} \tilde{f}_i(d)$, where the d's are dyadic. Clearly, f_i is a right continuous flow since $\tilde{\mathcal{A}}(u)$ is closed under intersection (cf. Lemma 1.1.3). We have $V_{f_i(u)} = \lim_{d \downarrow u} V_{\tilde{f}_i(d)} = V_{D_{i-1}} + u(V_{D_i} - V_{D_{i-1}})$ since V can be regarded as a measure on the σ-algebra generated by $\mathcal{A}$, and so $V \circ f_i$ is a continuous function.

Now let $f(s) = f_i(s - i + 1)$, if $i - 1 \le s \le i$, $s \in [0, k_1]$. Clearly f is a right-continuous flow and $f(i) = f_i(1) = f_{i+1}(0) = D_i$. Thus 1 and 2 are satisfied. Finally, 3 follows since for each i, $V \circ f_i$ is continuous. $\square$

5.2 A Martingale Characterization of Brownian Motion

We have the following characterization of the set-indexed Brownian motion from Ivanoff and Merzbach (1996):

Theorem 5.2.1 *Let X be a square-integrable $\mathcal{A}$-indexed process with $X_{\emptyset'} = 0$ which is monotone inner- and outer-continuous on $\mathcal{A}$. Let $\{\mathcal{F}_A : A \in \mathcal{A}\}$ be the minimal filtration (or the minimal outer-continuous filtration) generated by X. Then X is a Brownian motion with variance measure Λ if and only if the following conditions are satisfied:*

1. *Λ is deterministic, increasing and monotone inner- and outer-continuous on $\mathcal{A}$ with $\Lambda_{\emptyset'} = 0$.*

2. *X is a strong martingale with respect to its minimal filtration.*

3. *Λ is a *-quadratic variation for X.*

Proof.
Proof of 'Only if': Assume X is a Brownian motion with variance measure Λ. Properties 2 and 3 have already been proven. Property 1 follows from the inner- and outer-continuity of X on $\mathcal{A}$ and uniform integrability.

Proof of 'If': As noted in the previous section, it is enough to consider the joint distribution of $(X_{C_0} = 0, X_{C_1}, ..., X_{C_k})$ where $\{C_0 = \emptyset', C_1, ..., C_k\}$ may be assumed to be the left-neighbourhoods of the finite sub-semilattice $\mathcal{A}_1$. Let f be a flow satisfying the conditions of Lemma 5.1.7 and let $Y_r = X_{f(r)}$ and $\mathcal{H}_r = \mathcal{F}_{f(r)}$, for $0 \leq r \leq k$. By Lemma 5.1.2, Y is a $\mathcal{H}$-martingale and by Lemma 5.1.7, both Y and $\lambda = \Lambda \circ f$ are continuous. As well, $Y^2 - \lambda$ is a $\mathcal{H}$-martingale since

$$\begin{aligned}
E[Y_r^2 - Y_s^2 \mid \mathcal{H}_s] &= E[(Y_r - Y_s)^2 - 2Y_s^2 + 2Y_r Y_s \mid \mathcal{H}_s] \\
&= E[E(X_{f(r)\backslash f(s)})^2 \mid \mathcal{G}^*_{f(r)\backslash f(s)}] \mid \mathcal{H}_s] \\
&\quad + 2Y_s E[Y_r - Y_s \mid \mathcal{H}_s] \\
&= \Lambda_{f(r)\backslash f(s)} \\
&= \lambda_r - \lambda_s.
\end{aligned}$$

Now, by Lévy's characterization, Y is a Gaussian process with independent increments with variance measure λ. Thus, since $X_{C_i} = X_{f(i)} - X_{f(i-1)} = Y_i - Y_{(i-1)}$, and $\Lambda_{C_i} = \Lambda_{f(i)} - \Lambda_{f(i-1)} = \lambda_i - \lambda_{(i-1)}$, $i = 1, \ldots, k$, the random variables $X_{C_1}, \ldots, X_{C_k}$ are in-

THE POISSON PROCESS   107

dependent mean 0 normal random variables with variances $\Lambda_{C_1}, ...,$ Λ_{C_k}, respectively. This completes the proof. □

We note that the conditions in the preceding theorem continue (trivially) to be sufficient when the filtration $\mathcal{F}$ is larger than the minimal filtration.

As was noted previously, our characterization is in fact more general than the Wong-Zakai characterization of planar Brownian motion. For the Wong-Zakai characterization (as presented in Zakai (1981)), our condition (3) in the preceding theorem is replaced by:
(3a) $X^2 - \Lambda$ is a martingale with respect to the minimal filtration $\{\mathcal{F}_A, A \in \mathcal{A}\}$, and
(3b) the filtration $\{\mathcal{F}_A, A \in \mathcal{A}\}$ satisfies (CI).

Clearly, the two sets of conditions (1), (2), (3) and (1), (2), (3a), (3b) are equivalent when $\{\mathcal{F}_A, A \in \mathcal{A}\}$ is the minimal filtration. However, although both sets of conditions continue (trivially) to be sufficient when $\{\mathcal{F}_A, A \in \mathcal{A}\}$ is a larger filtration, they are no longer necessarily equivalent, as such a filtration may satisfy (3), but not (3b). (For example, if the minimal filtration is denoted by $\{\mathcal{F}'_A, A \in \mathcal{A}\}$ and if $\mathcal{F}''$ is a filtration independent of $\mathcal{F}'$ not satisfying (3b), then let $\mathcal{F}_A = \mathcal{F}'_A \vee \mathcal{F}''_A$.) This can be useful in applications in which it may be desirable to use a larger filtration.

5.3 A Martingale Characterization of the Poisson Process

In this section we give a martingale characterization of the Poisson process, proven in Ivanoff and Merzbach (1994). First, we recall that a simple point process N is an integer-valued increasing process on $\mathcal{A}$ such that $N_{\{t\}} = 0$ or 1 almost surely.

Theorem 5.3.1 *Let N be a simple $\mathcal{A}$-indexed point process and let $\mathcal{F} = \{\mathcal{F}_A = \mathcal{F}^r_A : A \in \mathcal{A}\}$ be its minimal outer-continuous filtration. Suppose that Λ is a monotone outer-continuous increasing function on $\mathcal{A}$ such that $\Lambda_{\{t\}} = 0$ for every $t \in T$. Then N is a Poisson process with mean measure Λ if and only if Λ is a *-compensator for N.*

Proof.
Proof of 'Only if': The 'only if' portion of the theorem is Theorem 4.5.2.
Proof of 'If': As in the case of the Brownian motion, without loss of generality it is enough to consider the joint distribution

of $(N_{C_0} = 0, N_{C_1}, ..., N_{C_k})$ where $\{C_0 = \emptyset', C_1, ..., C_k\}$ are the left-neighbourhoods of the finite sublattice $\mathcal{A}_1$. Let f be a flow satisfying the conditions of Lemma 5.1.9 and let $M_s = N_{f(s)}, \mathcal{H}_s = \mathcal{F}_{f(s)}$, for $0 \leq s \leq k$. Finally, let $\lambda = \Lambda \circ f$. By Lemma 5.1.2, $M - \lambda$ is a $\mathcal{H}$-martingale and by Lemma 5.1.9, λ is continuous. If it can be shown that M is a simple point process on $[0, k]$, then by Watanabe's characterization, it follows that Y is a Poisson process with mean measure λ. Thus, since $N_{C_i} = N_{f(i)} - N_{f(i-1)} = Y_i - Y_{i-1}$ and $\Lambda_{C_i} = \Lambda_{f(i)} - \Lambda_{f(i-1)} = \lambda_i - \lambda_{i-1}$, $i = 1, ..., k$, the random variables $N_{C_1}, ..., N_{C_k}$ are independent Poisson random variables with means $\Lambda_1, ..., \Lambda_k$, respectively.

Thus it remains only to show that M is a simple point process. We fix $n \geq 0$, and define $S_0^n, ..., S_{2^n}^n$ as follows: $S_0^n = 0$, $S_{2^n}^n = k$, and if $1 \leq i \leq 2^n - 1$, $S_i^n = \inf\{s \geq S_{i-1}^n : \lambda_s - \lambda_{S_{i-1}^n} = \lambda_k/2^n\}$. Thus, $\lambda_{S_i^n} = i\lambda_k/2^n, i = 0, ..., 2^n$. Now, it is easily seen that

$$\{\omega : M(\omega) \text{ not simple}\}$$
$$= \{\omega : \exists s \in [0, k] \text{ such that } M_s - M_{s-} > 1\}$$
$$\subseteq \cap_n \mathcal{U}_n$$

where $\mathcal{U}_n = \cup_{i=1}^{2^n}\{M_{S_i^n} - M_{S_{i-1}^n} > 1\}$. We shall show that $P(\mathcal{U}_n) \to 0$ as $n \to \infty$. Now,

$$P(\mathcal{U}_n)$$
$$= \cup_{i=1}^{2^n}\{M_{S_i^n} - M_{S_{i-1}^n} > 1\}$$
$$= \sum_{i=1}^{2^n} P\left(\{M_{S_i^n} - M_{S_{i-1}^n} > 1\} \cap \cap_{j \leq i-1}\{M_{S_j^n} - M_{S_{j-1}^n} \leq 1\}\right)$$
$$= \sum_{i=1}^{2^n} P\left(\{M_{S_i^n} - M_{S_{i-1}^n} > 1\} \cap \mathcal{V}^n(i-1)\right), \quad (5.1)$$

where $\mathcal{V}^n(i-1) = \cap_{j \leq i-1}\{M_{S_j^n} - M_{S_{j-1}^n} \leq 1\}$. Note that $\mathcal{V}^n(i-1) \in \mathcal{H}_{S_{i-1}^n} = \mathcal{F}_{f(S_{i-1}^n)}$.

We partition the set $f(S_i^n) \setminus f(S_{i-1}^n)$. For $r > n$, let $\mathcal{A}_{n,r}$ be the class of all intersections of sets in $\mathcal{A}_r \cup \{f(S_i^n); 0 \leq i \leq 2^n\}$ and let $\mathcal{A}_{n,r}(u)$ be the class of (finite) unions of sets in $\mathcal{A}_{n,r}$. Let $\mathcal{L}_{n,r}$ be the class of 'left-neighbourhoods' of sets in $\mathcal{A}_{n,r}$:

$$\mathcal{L}_{n,r} = \{C : C = B \setminus \cup_{B' \in \mathcal{A}_{n,r}, B \not\subseteq B'} B', B \in \mathcal{A}_{n,r}\}.$$

Let $\mathcal{L}_{n,r}(0) = \{\emptyset'\}$. Define for $i = 1, ..., 2^n$,

$$\mathcal{L}_{n,r}(i) = \{C \in \mathcal{L}_{n,r} : C \subseteq f(S_i^n) \backslash f(S_{i-1}^n)\}.$$

(We note here that the sets in $\mathcal{A}_{n,r}$ are in $\tilde{\mathcal{A}}(u)$, and the sets in $\mathcal{L}_{n,r}$ are not necessarily in $\mathcal{C}$.)

Now, $f(S_i^n) \backslash f(S_{i-1}^n) = \cup_{C \in \mathcal{L}_{n,r}(i)} C$ and $\mathcal{L}_{n,r} = \cup_{i=0}^{2^n} \mathcal{L}_{n,r}(i)$, and these unions are disjoint. Any set $C \in \mathcal{L}_{n,r}(i)$ is of the form $A \backslash B$, $A \in \mathcal{A}_{n,r}$, $B \in \mathcal{A}_{n,r}(u)$. The sets of $\mathcal{L}_{n,r}(i)$ (say $C_1, ..., C_q$) may be ordered so that if $C_j = A_j \backslash B_j$ and $C_h = A_h \backslash B_h$ and $A_h \subset A_j$, then $h < j$. Let '$\prec$' denote the induced total order on $\mathcal{L}_{n,r}(i)$ (i.e., $C \prec C' \Leftrightarrow C = C_h, C' = C_j$ where $h < j$). For $C \in \mathcal{L}_{n,r}(i)$, let $E_C = \cup_{C' : C \prec C'} C'$. (In some sense, E_C is the 'future' of C.) It is easily seen that E_C has the form $E_C = f(S_i^n) \backslash B_C$ for some $B_C \in \tilde{\mathcal{A}}(u)$ such that $f(S_{i-1}^n) \subseteq B_C$ and $C' \subseteq B_C$ if $C' \prec C$. Define:

$$\mathcal{D}_C = \{N_C = 1\} \cap \cap_{C' \prec C} \{N_{C'} = 0\}$$
$$\mathcal{E}_C = \{N_{E_C} \geq 1\}$$
$$\mathcal{Q}_{n,r}(i) = \{\omega : \exists C \in \mathcal{L}_{n,r}(i) \text{ such that } N_C > 1\}.$$

We observe that $\mathcal{D}_C \in \mathcal{F}(B_C)$, and that the sets $\mathcal{D}_C, \mathcal{D}_{C'}$ are disjoint if $C \neq C'$.

Now, if $N_{f(S_i^n) \backslash f(S_{i-1}^n)}(\omega) > 1$, then either one of the sets in $\mathcal{L}_{n,r}(i)$ contains more than one point, in which case $\omega \in \mathcal{Q}_{n,r}(i)$, or there exist two sets $C, C' \in \mathcal{L}_{n,r}(i)$ with $C \prec C'$ such that $N_C = 1$ and $N_{C'} = 1$, in which case $\omega \in \cup_{C \in \mathcal{L}_{n,r}(i)} (\mathcal{D}_C \cap \mathcal{E}_C)$. Thus,

$$\{M_{S_i^n} - M_{S_{i-1}^n} > 1\}$$
$$= \{N_{f(S_i^n) \backslash f(S_{i-1}^n)} > 1\}$$
$$\subseteq \{\cup_{C \in \mathcal{L}_{n,r}(i)} (\mathcal{D}_C \cap \mathcal{E}_C) \cup \mathcal{Q}_{n,r}(i)\}.$$

Therefore,

$$P\left(\{M_{S_i^n} - M_{S_{i-1}^n} > 1\} \cap \mathcal{V}^n(i-1)\right)$$
$$\leq \sum_{C \in \mathcal{L}_{n,r}(i)} P(\mathcal{D}_C \cap \mathcal{E}_C \cap \mathcal{V}^n(i-1)) + P(\mathcal{Q}_{n,r}(i))$$
$$= \sum_{C \in \mathcal{L}_{n,r}(i)} P(\mathcal{E}_C \mid \mathcal{D}_C \cap \mathcal{V}^n(i-1)) P(\mathcal{D}_C \cap \mathcal{V}^n(i-1))$$
$$+ P(\mathcal{Q}_{n,r}(i)).$$

Since $\mathcal{D}_C \cap \mathcal{V}^n(i-1) \in \mathcal{F}_{B_C}$, we may apply Comment 5.1.3.2 to show

$$P(\mathcal{E}_C \mid \mathcal{D}_C \cap \mathcal{V}^n(i-1)) \leq E[N_{E_C} \mid \mathcal{D}_C \cap \mathcal{V}^n(i-1)]$$
$$= E[\Lambda_{E_C} \mid \mathcal{D}_C \cap \mathcal{V}^n(i-1)]$$
$$= \Lambda_{E_C}.$$

Thus,

$$P\left(\{M_{S_i^n} - M_{S_{i-1}^n} > 1\} \cap \mathcal{V}^n(i-1)\right)$$
$$\leq \sum_{C \in \mathcal{L}_{n,r}(i)} \Lambda_{E_C} P(\mathcal{D}_C \cap \mathcal{V}^n(i-1))) + P(\mathcal{Q}_{n,r}(i))$$
$$\leq \frac{\lambda_k}{2^n} P(\mathcal{V}^n(i-1)) + P(\mathcal{Q}_{n,r}(i)). \qquad (5.2)$$

The last inequality above follows by the fact that $E_C \subseteq f(S_i^n) \setminus f(S_{i-1}^n)$, and the events $\mathcal{D}_C$ are disjoint.

Combining Equations (5.1) and (5.2) and noting that the events $\mathcal{V}_i^n$ and $\mathcal{V}_j^n$ are disjoint if $i \neq j$,

$$P\left(\cup_{i=1}^{2^n} \{M_{S_i^n} - M_{S_{i-1}^n} > 1\}\right)$$
$$= \sum_{i=1}^{2^n} P\left(\{M_{S_i^n} - M_{S_{i-1}^n} > 1\} \cap \mathcal{V}^n(i-1)\right)$$
$$\leq \frac{\lambda_k}{2^n} \sum_{i=1}^{2^n} P(\mathcal{V}^n(i-1)) + \sum_{i=1}^{2^n} P(\mathcal{Q}_{n,r}(i))$$
$$\leq \frac{\lambda_k}{2^n} + \sum_{i=1}^{2^n} P(\mathcal{Q}_{n,r}(i)).$$

However, r is arbitrary, so we may let $r \to \infty$ in the above expression holding n fixed. Since $\{\mathcal{C}_r\}_r$ separates points, it follows that as $r \to \infty$,

$$P(\mathcal{Q}_{n,r}(i)) \to P\{N \text{ is not simple on } f(S_i^n) \setminus f(S_{i-1}^n)\} = 0.$$

Therefore,

$$P(\mathcal{U}_n) = P\left(\cup_{i=1}^{2^n} \{M_{S_i^n} - M_{S_{i-1}^n} > 1\}\right) \leq \frac{\lambda_k}{2^n}.$$

Finally, we let $n \to \infty$ to show that M is a simple point process with probability one: $P\{\omega : M(\omega) \text{ not simple}\} = \lim_n P(\mathcal{U}_n) = 0$.

This completes the proof of the theorem. □

We observe that for a Poisson process on $\mathbf{R}_+^2$, it is possible that Λ is concentrated on a single line, in which case all of the points of the process will be restricted to this line. Therefore we do not require that there be no lines parallel to the axes with multiple jumps, we do not restrict $\mathcal{A}$ to be the class of rectangles, nor do we require (CI). These conditions were imposed in the previous characterizations of the planar Poisson process.

A similar characterization may be developed for the Cox process, using the same arguments as in Brémaud (1981).

CHAPTER 6

Generalizations of Martingales

In this chapter we extend the definitions of the different kinds of martingales to local martingales and to quasimartingales.

The concept of localization plays a fundamental role in the theory of stochastic processes. The aim of the first section is to introduce this notion in the framework of set-indexed processes, and to study the important class of processes called local martingales. As pointed out by P. A. Meyer (1981), even for two-parameter processes, localization is not an easy problem and a counterexample given by P. Imkeller (1986b) shows that 'bounded localization' is not possible in general. Despite these limitations, there are important classes of processes which can be localized in a suitable manner, and so it is essential that this concept be better understood.

The different types of local martingales are defined. We also introduce the class of processes which are locally of class D. Notice that in the set-indexed framework, the class D condition is very delicate. For example, even a closable martingale is not necessarily of class D. As an application, we extend the Doob-Meyer decomposition to local weak submartingales and show the existence of a unique decomposition into the sum of a local weak martingale and a predictable increasing process. Since a weak martingale may have very bad regularity properties, we cannot expect to obtain better results. Finally, non-trivial examples of such local weak submartingales are presented. The material presented in the two first sections appears in Ivanoff and Merzbach (1995).

The last section is devoted to quasimartingales. This notion was defined first in Dozzi, Ivanoff and Merzbach (1994) and investigated further by Burstein (1999).

Although of independent interest, this chapter is not required for the sequel and may be omitted on first reading.

6.1 Local Martingales

Definition 6.1.1 *Let $\mathcal{E}$ be a class of processes and X be an arbitrary process, which is not necessarily integrable. We say that a stopping set ξ reduces the process X with respect to the class $\mathcal{E}$ if $X^\xi = \{X_{A\cap\xi}, A \in \mathcal{A}\} \in \mathcal{E}$ and X^ξ is of class D.*

We say that X has a localizing sequence $\{\xi_n\}$ *with respect to the class $\mathcal{E}$ if $\{\xi_n\}$ is an increasing sequence of stopping sets which reduce X with respect to $\mathcal{E}$ and such that $\cup_{n=1}^\infty (\xi_n(\omega))^\circ = T$, for all $\omega \in \Omega$.*

We are now in position to define local martingales. In the two-parameter setting, a similar definition was proposed by A. A. Gushchin and Y. S. Mishura (1992) for strong locally square integrable martingales, and another one was suggested by J.P. Fouque (1983).

Definition 6.1.2 *A process X is called a* local (strong, weak) (sub-)martingale *if it has a localizing sequence with respect to the class of (strong, weak) (sub-)martingales.*

Remark 6.1.3 Notice that a local (strong, weak) martingale of class D is a (strong, weak) martingale. Also, any local (strong, weak) (sub-)martingale whose localizing sequence is deterministic is a (strong, weak) (sub-)martingale.

The following notation will be used in the sequel:
$\sum_X(\mathcal{E}) = \{\xi : \xi \text{ reduces } X \text{ with respect to } \mathcal{E}\}$.

Definition 6.1.4 *We say that the process X is* locally of class D *if it has a localizing sequence with respect to the class of all processes (i.e., there exists an increasing sequence $\{\xi_n\}$ of s.s.'s such that $\cup_{n=1}^\infty (\xi_n(\omega))^\circ = T$, for all $\omega \in \Omega$ and X^{ξ_n} is of class D, $\forall n$.)*

Remark 6.1.5 A (strong, weak) martingale locally of class D is a local (strong, weak) martingale. This fact follows easily from Theorem 3.3.4 (resp. Theorem 3.3.1 and Theorem 3.3.8).

The following three results are from Ivanoff and Merzbach (1995).

Proposition 6.1.6 *The class $\sum_X(\mathcal{E})$ is closed under finite intersections, and if SHAPE holds, under finite unions, where $\mathcal{E}$ is either the class of (strong, weak) martingales, or $\mathcal{E}$ is the class of processes which are locally of class D.*

Proof. Let ξ and ξ' be two stopping sets which reduce X with respect to $\mathcal{E}$. We have to show that the same holds for the stopping set $\xi \cap \xi'$ and for the stopping set $\xi \cup \xi'$ (SHAPE is needed to ensure that $\xi \cup \xi'$ is a s.s.). By the equality $X^{\xi \cup \xi'} = X^{\xi} + X^{\xi'} - X^{\xi \cap \xi'}$, it follows that it is enough to prove that $\xi \cap \xi'$ reduces X (indeed, using the admissible function μ_X, if the processes $X^{\xi}, X^{\xi'}$ and $X^{\xi \cap \xi'}$ are weak (or strong) martingales, then $X^{\xi \cup \xi'}$ will have the same property).

Now, begin with the case 'locally of class D' : since $\xi \cap \xi' \subseteq \xi$ and $\xi \cap \xi' \subseteq \xi'$ then if X^{ξ} and $X^{\xi'}$ are both of class D then $X^{\xi \cap \xi'}$ is also of class D. Therefore the other cases now follow easily as before from Theorem 3.3.4, Theorem 3.3.1 and Theorem 3.3.8. □

This proposition is not necessarily true for submartingales: Indeed, the fact that X^{ξ} and $X^{\xi'}$ are submartingales does not imply that $X^{\xi \cup \xi'}$ is a submartingale. However $\sum_X (\mathcal{E})$ is closed under finite intersections if $\mathcal{E}$ is the class of (strong, weak) submartingales.

Corollary 6.1.7 *The sum of two local (strong, weak, strong sub-, weak sub-, sub-)martingales is still a local (strong, weak, strong sub-, weak sub-, sub-)martingale.*

Proof. Let X and Y be two processes which belong to one of these classes, and $\{\xi_n\}, \{\xi'_n\}$ their respective localizing sequences. Following Proposition 6.1.6, $\{\xi_n \cap \xi'_n\}$ is a localizing sequence for each of them. Therefore, it is a localizing sequence for $X + Y$. □

Proposition 6.1.8 *Assume SHAPE. Then a nonnegative local martingale of class D is a supermartingale.*

Proof. Let $\{\xi_n\}$ be a localizing sequence. Since $\cup_{n=1}^{\infty} (\xi_n(\omega))^{\circ} = T$, for each $A \in \mathcal{A}$, $\{A \subseteq \xi_n\} \subseteq \{A \subseteq \xi_{n+1}\} \uparrow \Omega$.

Let $F \in \mathcal{F}_A$ for $A \subseteq B$.

$$\int_F X_B dP = \lim_{n \to \infty} \int_{F \cap \{B \subseteq \xi_n\}} X_B dP = \lim_{n \to \infty} \int_{F \cap \{B \subseteq \xi_n\}} X_B^{\xi_n} dP$$

(since $X_B = X_B^{\xi_n}$ on $\{B \subseteq \xi_n\}$)

$$\leq \lim_{n \to \infty} \int_{F \cap \{A \subseteq \xi_n\}} X_B^{\xi_n} dP$$

(since $\{B \subseteq \xi_n\} \subseteq \{A \subseteq \xi_n\}$ and $X_B \geq 0$)

$$= \lim_{n \to \infty} \int_{F \cap \{A \subseteq \xi_n\}} X_A^{\xi_n} dP$$

(since $\{X_A^{\xi_n}, \mathcal{F}_A, A \in \mathcal{A}\}$ is a martingale)

$$= \lim_{n \to \infty} \int_{F \cap \{A \subseteq \xi_n\}} X_A dP = \int_F X_A dP. \quad \square$$

6.2 Doob-Meyer Decompositions

We now give an important application of the concept of localization, by extending the Doob-Meyer decomposition to weak submartingales not necessarily of class D. A similar result can be obtained for strong submartingales. The results in this section are from Ivanoff and Merzbach (1995).

Theorem 6.2.1 *Suppose that Assumption 4.3.2 holds and let X be a local weak submartingale. Then there exists a unique decomposition $X = M + V$, where M is a local weak martingale and V is a local predictable increasing process.*

Proof. Let $\{\xi_n\}$ be a localizing sequence for X. Then, for each n, we have a unique decomposition: $X^{\xi_n} = M^n + V^n$, where M^n is a weak submartingale and V^n a predictable increasing process (Corollary 4.3.7(i)). Note that $M^n_{A\setminus \xi_n} = V^n_{A\setminus \xi_n} = 0, \forall A \in \mathcal{A}$. If $F \times C \subseteq (\phi, \xi_n]$, with $F \in \mathcal{G}_c$, then

$$\mu_X(F \times C) = E(I_F X_C) = E(I_F X_{C \cap \xi_n}) = \mu_{X^{\xi_n}}(F \times C).$$

Therefore on $(\phi, \xi_n]$, we have $\mu_X = \mu_{X^{\xi_n}}$. Note that V^n is of class D since $0 \leq V^n_\tau \leq V^n_{\xi_n}$, which is integrable. Thus, M^n is also of class D. We now show that $(V^{n+1})^{\xi_n} = V^n$. Following Theorem 4.3.6(i), it is enough to check that the two vector measures $\nu^{\xi_n}_{n+1}$ and ν_n associated with these two processes coincide. Now, coming back to the proof of Theorem 4.3.6(i), we notice that the construction of ν' is exactly the same for both the processes X^{ξ_n} and $X^{\xi_{n+1} \cap \xi_n}$. Therefore $(V^{n+1})^{\xi_n} = V^n$ and this is a predictable increasing process.

Now define the processes M and V as follows:

$$M_A(\omega) = M^n_A(\omega) \text{ and } V_A(\omega) = V^n_A(\omega) \text{ if } A \subseteq \xi_n(\omega).$$

Note: $M^{\xi_n}_A = M_{A \cap \xi_n} = M^n_{A \cap \xi_n} = M^n_A$. Thus M is a local martingale. The process V is increasing and so has a σ-additive admissible measure μ_V on $\mathcal{P}$. Also, on $(\phi, \xi_n]$, $\mu_V = \mu_{V^{\xi_n}}$ (Same argument as before).

To show that V is predictable, consider any rectangle $F \times C$ with $F \in \mathcal{F}$ and $C \in \mathcal{C}$ and let $C = A \setminus \cup_{i=1}^k A_i$. Note that $\{\omega : C \subseteq \xi_n(\omega)\} \subseteq H_n$, where $H_n = \{\omega : A \subseteq \xi_n(\omega)\} \in \mathcal{F}_A$. Denote $\Omega_1 = H_1, \Omega_n = H_n \setminus H_{n-1}, F_n = F \cap \Omega_n$. Then, clearly $\Omega = \cup_{n=1}^\infty \Omega_n, F = \cup_{n=1}^\infty F_n$, where $F_n \in \mathcal{F}$ for any n, and the sets $\{F_n\}$

DOOB-MEYER DECOMPOSITIONS   117

are disjoint. Thus,

$$
\begin{aligned}
\mu_V(F \times C) &= \sum_{n=1}^{\infty} \mu_V(F_n \times C) \\
&= \sum_{n=1}^{\infty} \mu_{V^{\xi_n}}(F_n \times C) \text{ since } F_n \times C \subseteq (\phi, \xi_n] \\
&= \sum_{n=1}^{\infty} \int E(I_{F_n}|\mathcal{F}_t) I_C(t) d\mu_{V^{\xi_n}} \\
&= \sum_{n=1}^{\infty} \int E(I_{F_n}|\mathcal{F}_t) I_C(t) d\mu_V \\
&= \int E(I_F|\mathcal{F}_t) I_C(t) d\mu_V,
\end{aligned}
$$

and therefore V is predictable.

By the same arguments, using the uniqueness in Theorem 4.3.6(i), we obtain the uniqueness of the decomposition. $\square$

Corollary 6.2.2 *If Assumption 4.3.2 holds, then for any increasing process X there exists a unique decomposition $X = M + V$, where M is a local weak martingale and V is a local predictable increasing process.*

Proof. This follows immediately from Theorem 6.2.1 since an increasing process is a weak submartingale locally of class D and the localizing sequence may be chosen to be deterministic. $\square$

We now give two examples. The first is a non-trivial weak submartingale on $T = \mathbf{R}_+^2$ which is not necessarily uniformly integrable, but is locally of class D. The second is a non-integrable point process on $T = \mathbf{R}^d$. In both cases, we let $\mathcal{A} = \{A_z : z \in T\}$ where $A_z = [0, z]$.

Examples 6.2.3 (1) Let $T = \mathbf{R}_+^2$ and let $\{U_s, \mathcal{F}_s^U\}, \{V_r, \mathcal{F}_r^V\}$ be independent submartingales on $\mathbf{R}_+$ with $U_0 = V_0 = 0$. Let $\{X, \mathcal{F}\}$ be defined on $\mathcal{A}$ as follows:

$$
\begin{aligned}
X_{A_{s,r}} &= U_s V_r, \\
\mathcal{F}_{A_{s,r}} &= \mathcal{F}_s^U \vee \mathcal{F}_r^V.
\end{aligned}
$$

It is easy to see that on any rectangle $C = (s, s'] \times (r, r']$

$$
E(X_C|\mathcal{G}_C) = E((U_{s'} - U_s)(V_{r'} - V_r)|\mathcal{F}_{A_{(s,r)}})
$$

$$= E(U_{s'} - U_s|\mathcal{F}^U_s)E(V_{r'} - V_r|\mathcal{F}^V_r)$$
$$\geq 0.$$

It is then straightforward to verify that $E(X_C|\mathcal{G}_C) \geq 0$ for any set C in $\mathcal{C}$, so X is a weak submartingale. (In fact it is also a submartingale.)

Now suppose that $\mathcal{F}^U$ and $\mathcal{F}^V$ are both right-continuous, and let S be a bounded stopping time with respect to $\mathcal{F}^U$ and R a bounded stopping time with respect to $\mathcal{F}^V$. By right-continuity of the filtrations, $\{s \leq S\} \in \mathcal{F}^U_s$ and $\{r \leq R\} \in \mathcal{F}^V_r$. Thus, $A_{S,R}$ is a simple stopping set, since for $\forall(s,r) \in \mathbf{R}^2_+$,

$$\{A_{s,r} \subseteq A_{S,R}\} = \{s \leq S\} \cap \{r \leq R\} \in \mathcal{F}_{A_{s,r}}.$$

Suppose that one of the submartingales, say V, is increasing. In this case, we claim that X is locally of class D. Indeed, let

$$S_n = \inf\{s : U_s \geq n\} \wedge n$$
$$R_n = \inf\{r : V_r \geq n\} \wedge n.$$

We have that both U_{S_n} and V_{R_n} are integrable, and if $U^{S_n}_s = U(S_n \wedge s)$, then $|U^{S_n}_s| \leq n \vee |U_{S_n}|$. Likewise $|V^{R_n}_r| \leq n \vee |V_{R_n}|$.

Clearly, the random sets $\{\xi_n\} = \{A_{S_n, R_n}\}$ converge to $\mathbf{R}^2_+$. We shall show that X^{ξ_n} is of class D. Fix ω, and let $A = \xi_n(\omega)$. Let $B \in \mathcal{A}(u)$. Then $A \cap B$ may be expressed as a finite disjoint union of horizontal bands $C_1, \ldots, C_k$ where $C_1 = [0, s_1] \times [0, r_1]$, and for $2 \leq i \leq k, C_i = [0, s_i] \times (r_{i-1}, r_i]$, where $0 \leq s_k < s_{k-1} < \ldots < s_1 \leq S_n(\omega)$ and $0 \leq r_1 < r_2 < \ldots < r_k \leq R_n(\omega)$. We have

$$X_{A \cap B} = \sum_{i=1}^k X_{C_i}$$
$$= \sum_{i=1}^k U_{s_i} \times (V_{r_i} - V_{r_{i-1}}) \text{ (setting } r_0 = 0\text{).}$$

Hence

$$|X_{A \cap B}| \leq \sum_{i=1}^k |U_{s_i}|(V_{r_i} - V_{r_{i-1}}), \text{ since } V \text{ is increasing}$$
$$\leq |n \vee U_{S_n}| \times V_{r_k}$$
$$\leq |n \vee U_{S_n}||n \vee V_{R_n}|.$$

It follows that for any stopping set τ,
$$|X_\tau^{\xi_n}(\omega)| \leq n^2 + n|U_{S_n}(\omega)| + nV_{R_n}(\omega) + |U_{S_n}(\omega)V_{R_n}(\omega)|.$$
Since the right hand side is integrable, X^{ξ_n} is of class D. It is easily verified that if $\hat{U}$ and $\hat{V}$ are the (one-dimensional) dual predictable projections of U and V, respectively, then the local predictable increasing process in the Doob-Meyer decomposition is $\hat{U}\hat{V}$.

(2) Let $T = \mathbf{R}^d$ and $\mathcal{A} = \{\prod_{i=1}^d [0, t_i] : 0 \leq |t_i| < \infty\}$. Let X be a $\mathcal{F}_0$-measurable positive and non-integrable random variable strictly bounded away from 0 (i.e., there exists $\epsilon > 0$ such that $P(X \geq \epsilon) = 1$). Define a point process N on T as follows: given $\{X = x\}$, N is a Poisson process on T with mean measure $x\lambda$, where λ is Lebesgue measure (i.e., N is a Cox process with driving measure $X\lambda$). It is easily seen that N is a local strong submartingale with localizing sequence $[-n/X, n/X]$, where $n/X = (n/X, ..., n/X)$. The local predictable increasing process in the Doob-Meyer decomposition is $X\lambda$.

6.3 Quasimartingales

In the two-parameter case, several authors studied the notion of quasimartingale and tried to define two-parameter semimartingales. In particular, we refer the reader to Stoica (1978), Föllmer (1979), Brennan (1979), Guyon and Prum (1981), Bakry (1982), Merzbach and Zakai (1986), and Dozzi (1989).

Set-indexed quasimartingales were defined for the first time in Dozzi, Ivanoff and Merzbach (1994) and extensively studied in Burstein (1999). With the exception of Theorem 6.3.8, all the results in this section are from Burstein (1999), although in some cases the proofs here have been simplified.

Definition 6.3.1 *Let $X = \{X_A, A \in \mathcal{A}\}$ be an integrable, adapted and L^1-monotone outer-continuous process.*

X is called a weak *(resp.* strong*) quasimartingale if for all $A \in \mathcal{A}, A \neq \emptyset$,*
$$Q^{(*)}(X, A) = \sup E \sum_{i=1}^n |E(X_{C_i} \mid \mathcal{G}_{C_i}^{(*)})| < \infty, \quad (6.1)$$
where the supremum is taken over all the finite partitions $\{C_1, \ldots, C_n\}$ of A by sets of $\mathcal{C}$.

Let $\Pi(A)$ be the collection of all finite increasing sequences

$(A_i)_{i=1}^n$ of sets in $\mathcal{A}$, such that $\cup_{i=1}^n A_i = A$, $A_0 = \emptyset$. Define

$$\tilde{Q}(X, A) = \sup_{\Pi(A)} E \sum_{i=1}^n |E(X_{A_i} - X_{A_{i-1}} \mid \mathcal{F}_{A_{i-1}})|, \qquad (6.2)$$

and denote $\tilde{Q}(X) = \tilde{Q}(X, T)$, if $T \in \mathcal{A}$.

X is called a quasimartingale if and only if $\tilde{Q}(X, A) < +\infty$, for any $A \in \mathcal{A}$.

Remarks 6.3.2 1. Any L^1-monotone outer-continuous (weak, strong) submartingale or supermartingale is a (weak, strong) quasimartingale.

2. The sum and the difference of two weak (strong) quasimartingales is a weak (strong) quasimartingale.

3. Let X be an adapted process. Then $\tilde{Q}(X, A) = 0$ for all $A \in \mathcal{A} \Leftrightarrow X$ is a martingale. If moreover $T \in \mathcal{A}$, then $\tilde{Q}(X) = 0 \Leftrightarrow X$ is a martingale.

4. X is a weak (strong) quasimartingale is equivalent to the condition that μ_X is of bounded variation on $r(\mathcal{P}^{(*)}) \cap (\Omega \times A)$ for each $A \in \mathcal{A}$; and in this case $|\mu_X|(\Omega \times A) = Q^{(*)}(X, A)$ is the total variation of μ_X on $\Omega \times A$.

5. If X is a quasimartingale, then for any infinite bounded increasing sequence (A_n) in $\mathcal{A}$, $E \sum_{n=1}^\infty |E(X_{A_n} - X_{A_{n-1}} | \mathcal{F}_{A_{n-1}})| < \infty$.

Proposition 6.3.3 *If X is a strong quasimartingale then it is a quasimartingale.*

Proof. Let X be a strong quasimartingale and let $A \in \mathcal{A}$. For any increasing sequence (A_i) in $\Pi(A)$, set $C_i = A_i \setminus A_{i-1}, \forall i$. We have $\mathcal{F}_{A_{i-1}} \subseteq \mathcal{G}^*_{C_i}$, and therefore

$$\begin{aligned}
\tilde{Q}(X, A) &= \sup_{\Pi(A)} E \sum_{i=1}^n |E(X_{A_i} - X_{A_{i-1}} \mid \mathcal{F}_{A_{i-1}})| \\
&\leq \sup_{\substack{C_i = A_i \setminus A_{i-1} \\ \{A_i\} \in \Pi(A)}} E \sum_{i=1}^n |E(X_{C_i} \mid \mathcal{G}^*_{C_i})| \\
&\leq \sup E \sum_{i=1}^n |E(X_{C_i} \mid \mathcal{G}^*_{C_i})| = Q^*(X, A) < \infty
\end{aligned}$$

by (6.1). □

Recall (Definition 3.1.6) that a process X is $\mathcal{A}$-L^2-bounded if $\sup_{A \in \mathcal{A}} E|X_A^2| < \infty$.

QUASIMARTINGALES 121

Proposition 6.3.4 *Assume SHAPE and (CI). Let X and Y be two $\mathcal{A}$-L^2-bounded L^1-monotone outer-continuous martingales. Then the process*

$$X \cdot Y = \{X_A \cdot Y_A, A \in \mathcal{A}\}$$

is both a quasimartingale and a weak quasimartingale.

Proof. Fix $A \in \mathcal{A}$. For any finite increasing sequence $(A_i)_{i=1}^n$ of sets in $\mathcal{A}$, such that $\cup_{i=1}^n A_i = A, A_0 = \emptyset$ we have

$$\sum_{i=1}^n E|E(X_{A_i}Y_{A_i} - X_{A_{i-1}}Y_{A_{i-1}} \mid \mathcal{F}_{A_{i-1}})|$$

$$= \sum_{i=1}^n E|E((X_{A_i} - X_{A_{i-1}})(Y_{A_i} - Y_{A_{i-1}}) \mid \mathcal{F}_{A_{i-1}})|$$

$$\leq \sum_{i=1}^n E|(X_{A_i} - X_{A_{i-1}})(Y_{A_i} - Y_{A_{i-1}})|$$

$$\leq \sum_{i=1}^n \|X_{A_i} - X_{A_{i-1}}\|_2 \|Y_{A_i} - Y_{A_{i-1}}\|_2$$

(by Hölder's inequality)

$$\leq \sum_{i=1}^n \|X_{A_i} - X_{A_{i-1}}\|_2 \sum_{i=1}^n \|Y_{A_i} - Y_{A_{i-1}}\|_2$$

$$\leq (\sum_{i=1}^n \|X_{A_i} - X_{A_{i-1}}\|_2^2)^{\frac{1}{2}} (\sum_{i=1}^n \|Y_{A_i} - Y_{A_{i-1}}\|_2^2)^{\frac{1}{2}}. \quad (6.3)$$

Since X, Y are martingales, $\|X_{A_i} - X_{A_{i-1}}\|_2^2 = E(X_{A_i}^2 - X_{A_{i-1}}^2)$ for each i, and similarly for Y. So the final expression in (6.3) is equal to

$$(\sum_{i=1}^n E(X_{A_i}^2 - X_{A_{i-1}}^2))^{\frac{1}{2}} (\sum_{i=1}^n E(Y_{A_i}^2 - Y_{A_{i-1}}^2))^{\frac{1}{2}},$$

which telescopes to

$$(E(X_A^2 - X_\emptyset^2))^{\frac{1}{2}} (E(Y_A^2 - Y_\emptyset^2))^{\frac{1}{2}} = \|X_A\|_2 \|Y_A\|_2$$
$$\leq \|X\|_2 \|Y\|_2,$$

where $\|Y\|_2 = \sup_{A \in \mathcal{A}} \|Y_A\|_2$. This shows that XY is a quasimartingale.

For the weak case, using the fact that

$$X \cdot Y = \frac{1}{4}[(X+Y)^2 - (X-Y)^2],$$

we have only to note that under (CI) and SHAPE the square of a square integrable martingale is a weak submartingale (Proposition 3.1.8) and to use Proposition 6.3.8 following. □

Another non-trivial example of a quasimartingale is the following:

Proposition 6.3.5 *Let $\{X_A, A \in \mathcal{A}\}$ be a monotone outer-continuous adapted and integrable process and let $\{\mathcal{F}_A^0, A \in \mathcal{A}\}$ be its natural filtration ($\mathcal{F}_A^0 = \sigma(\{X_B, B \subseteq A, B \in \mathcal{A}\})$). Suppose that for every $A, B \in \mathcal{A}$, $A \subseteq B$, $X_{B \setminus A}$ is independent of $\mathcal{F}_A^0$. (This can be regarded as an extension of the notion of a process with independent increments.) If the function $A \to E(X_A)$ is of bounded variation, then X is a quasimartingale, which can be decomposed into the sum of a martingale and a process of bounded variation.*

Proof. For $A \subseteq B$ we have $E(X_B \mid \mathcal{F}_A^0) = E(X_B - X_A + X_A \mid \mathcal{F}_A^0) = E(X_{B \setminus A} \mid \mathcal{F}_A^0) + X_A = E(X_B - X_A) + X_A$, so if $B \to E(X_B)$ is of bounded variation, then X is a quasimartingale. In this case, set $V_B = E(X_B)$. So $V = \{V_A, A \in \mathcal{A}\}$ is a deterministic process with bounded variation, and $X - V$ is a martingale. Indeed, $E(X_B - V_B \mid \mathcal{F}_A^0) = E(X_B - X_A + X_A - V_B \mid \mathcal{F}_A^0) = E(X_{B \setminus A} \mid \mathcal{F}_A^0) + X_A - E(E(X_B) \mid \mathcal{F}_A^0) = E(X_B - X_A) + X_A - E(X_B) = X_A - V_A$. □

Theorem 6.3.6 *If X is a weak quasimartingale of class D or D', then μ_X can be extended to a measure on $\mathcal{P}$.*

Proof. We know that μ_X is of bounded variation on $r(\mathcal{P}_0) \cap (\Omega \times A)$ for any $A \in \mathcal{A}$ and we need to prove that μ_X has a σ-additive extension to $\mathcal{P}$. As is noted in Métivier (1982), the existence of a bounded σ-additive extension for $|\mu_X|$ implies the existence of a σ-additive extension for μ_X. Thus, it suffices to show that $|\mu_X|$ has a σ-additive extension to $\mathcal{P}$.

As in Theorem 4.2.8, it is enough to show that for any sequence (R_n) of sets in $r(\mathcal{P}_0)$ such that $R_n \downarrow \emptyset$, $|\mu_X|(R_n) \downarrow 0$. The proof follows the main lines of the proof of Theorem 4.2.8, and so will not be repeated here. □

Since we do not have an analogue of class D for the strong case, we obtain only the following:

Proposition 6.3.7 *If X is a strong quasimartingale of class $(D')^*$, then μ_X can be extended to a measure on $\mathcal{P}^*$.*

Proof. The proof is similar to that of Theorem 6.3.6 for D' weak quasimartingales, replacing $\mathcal{P}$ by $\mathcal{P}^*$ and $\mathcal{G}$ by $\mathcal{G}^*$. □

The next proposition was proved in Dozzi, Ivanoff and Merzbach (1994).

Proposition 6.3.8 *Any difference of two L^1-monotone outer-continuous weak submartingales is a weak quasimartingale. Conversely, if X is a weak quasimartingale, then there exist two weak submartingales Y and Z such that for all $C \in \mathcal{C}$,*

$$\mu_X(F \times C) = \mu_Y(F \times C) - \mu_Z(F \times C) \quad \forall F \in \mathcal{G}_C.$$

Proof. X is a quasimartingale means that μ_X has bounded variation in $r(\mathcal{P}_0) \cap (\Omega \times A)$ for each $A \in \mathcal{A}$. However, this is equivalent to saying that

$$\mu_X = \mu^+ - \mu^-,$$

where μ^+ and μ^- are finitely additive and nonnegative on $r(\mathcal{P}_0)$.

Therefore, the first part follows immediately. Conversely, for $C \in \mathcal{C}$ fixed, let us define two processes Y and Z by

$$Y_C = \frac{d\mu^+(\cdot, C)}{dP} \quad \text{and} \quad Z_C = \frac{d\mu^-(\cdot, C)}{dP} \quad \text{on } \mathcal{G}_C.$$

Y_C and Z_C are well defined since μ^+ and μ^- are measures on $(\Omega, \mathcal{G}_C)$ for each $C \in \mathcal{C}$, absolutely continuous with respect to P.

Then for $F \in \mathcal{G}_C$, we have:

$$E[I_F Y_C] = \mu^+(F \times C) \quad \text{and} \quad E[I_F Z_C] = \mu^-(F \times C).$$

Since μ^+ and μ^- are nonnegative, this implies that Y and Z are submartingales. We also have that for $F \in \mathcal{G}_C$:

$$E[I_F X_C] = \mu_X(F \times C) = \mu^+(F \times C) - \mu^-(F \times C) = E[I_F(Y_C - Z_C)].$$

□

Comment. In the classical theory, any quasimartingale is the difference of two submartingales. But, as was noted in Dozzi, Ivanoff and Merzbach (1994), the same proof does not hold in the set-indexed case, since we cannot deal with the notion of a strict future of a set $C \in \mathcal{C}$. As well, it is not clear that Proposition 6.3.8 can be extended to strong quasimartingales, since conditioning with respect to the σ-algebras $\mathcal{G}_C^*$ will not yield an adapted process in general. As a consequence, we obtain only the weaker result of Proposition 6.3.8.

The following proposition can be proved as Corollary 4.3.7, and so we omit the proof.

Proposition 6.3.9 *Assume SHAPE and Assumption 4.3.2. Let X be a weak quasimartingale such that μ_X can be extended to a σ-additive measure on $\mathcal{P}$. (In particular, this is the case if the two weak submartingales of Proposition 6.3.8 are of class D or D'.) Then $X = M + V - V'$, where M is a weak martingale and V and V' are predictable increasing processes.*

For a weak quasimartingale, if $T \in \mathcal{A}$, the process $V - V'$ of Proposition 6.3.9 may be approximated directly, generalizing a result of Duc, Nualart and Sanz (1991). Assume that $T \in \mathcal{A}$. Let $\{C_i^{(n)}\}_{i=1}^{k(n)}$ be an increasing sequence of partitions of T by left-neighbourhoods from $\mathcal{C}^\ell(\mathcal{A}_n)$, and define for any $C \in \mathcal{C}$

$$\Delta^n(X)_C = \sum_{i=1}^{k(n)} |E(X_{C \cap C_i^{(n)}} \mid \mathcal{G}_{C \cap C_i^{(n)}})|,$$

$$\Delta^n(X)_C^+ = \sum_{i=1}^{k(n)} (E(X_{C \cap C_i^{(n)}} \mid \mathcal{G}_{C \cap C_i^{(n)}}))^+,$$

$$\Delta^n(X)_C^- = \sum_{i=1}^{k(n)} (E(X_{C \cap C_i^{(n)}} \mid \mathcal{G}_{C \cap C_i^{(n)}}))^-,$$

so that $\Delta^n(X)_C = \Delta^n(X)_C^+ + \Delta^n(X)_C^-$. Notice that for $n \leq m$, $E(\Delta^n(X)_C) \leq E(\Delta^m(X)_C)$. Let $Q(X)_A = Q(X, A)$. If X is a weak quasimartingale, by L^1-monotone outer-continuity of X, it is not difficult to show that the function $A \to Q(X)_A$ is monotone outer-continuous on $\mathcal{A}$. Define the sets of random variables $I_X(C) = \{\Delta^n(X)_C, n \geq 1\}, I_X(C)^+ = \{\Delta^n(X)_C^+, n \geq 1\}, I_X(C)^- = \{\Delta^n(X)_C^-, n \geq 1\}$, and $I_X = I_X(T) = \{\Delta^n(X)_T, n \geq 1\}$.

Theorem 6.3.10 *Assume SHAPE and that $T \in \mathcal{A}$. Let X be a monotone outer-continuous weak quasimartingale and suppose the set I_X is uniformly integrable. Then, there exists a (not necessarily unique) decomposition $X = M + V$, where M is a weak martingale and V is a process with paths of bounded variation.*

Proof. The proof of this result follows from the main lines of Duc,

Nualart and Sanz (1991). We first prove that a weak quasimartingale, when indexed by the restricted family $\mathcal{A}_0 = \cup_p \mathcal{A}_p$, can be decomposed into the sum of a weak martingale and a process of bounded variation, both indexed by $\mathcal{A}_0$. Then we extend to $\mathcal{A}$ by a continuity argument.

Let $\mathcal{C}_0 = \cup_p \mathcal{C}^\ell(\mathcal{A}_p)$ denote the class of left-neighbourhoods generated by the sub-semilattices $\mathcal{A}_p, p \geq 1$, and number the sets in $\mathcal{C}_0 = \{C_0 = \emptyset', C_1, C_2, ...\}$ in some manner. We show the existence of a subsequence $\{n^{(n)}, n \geq 1\} \in \mathbf{N}$, and integrable random variables $V^+(C_k), V^-(C_k), C_k \in \mathcal{C}_0$, such that

$$\Delta^{n^{(n)}}(X)^+_{C_k} \to V^+(C_k), \quad \Delta^{n^{(n)}}(X)^-_{C_k} \to V^-(C_k), \qquad (6.4)$$

respectively, in the weak topology $\sigma(L^1, L^\infty)$, as $n \to \infty, \forall k$.

Indeed, as the collections $I^+_X(C_1)$ and $I^-_X(C_1)$ are uniformly integrable, there exist a subsequence $\{\Delta^{n^{(1)}}, n \geq 1\}$ of $\{\Delta^n, n \geq 1\}$ and integrable random variables $V^+(C_1), V^-(C_1)$ such that

$$\Delta^{n^{(1)}}(X)^+_{C_1} \xrightarrow{\sigma(L^1, L^\infty)} V^+(C_1),$$
$$\Delta^{n^{(1)}}(X)^-_{C_1} \xrightarrow{\sigma(L^1, L^\infty)} V^-(C_1), \text{ as } n \to \infty.$$

In this way we can construct recursively a subsequence $\{\Delta^{n^{(i+1)}}, n \geq 1\}$ of $\{\Delta^{n^{(i)}}, n \geq 1\}$, and integrable random variables $V^+(C_{i+1}), V^-(C_{i+1}), i \geq 1$, such that

$$\Delta^{n^{(i+1)}}(X)^+_{C_{i+1}} \xrightarrow{\sigma(L^1, L^\infty)} V^+(C_{i+1}),$$
$$\Delta^{n^{(i+1)}}(X)^-_{C_{i+1}} \xrightarrow{\sigma(L^1, L^\infty)} V^-(C_{i+1}), \text{ as } n \to \infty.$$

The diagonal sequence $\{\Delta^{n^{(n)}}, n \geq 1\}$, and the integrable random variables $\{V^+(C_k), V^-(C_k), k \geq 1\}$ satisfy (6.4). Because the sequence $(\mathcal{C}^\ell(\mathcal{A}_p))_p$ is increasing, these limits have finitely additive sample paths on $\mathcal{C}_0$. Indeed, for any j, if $C_j = C_{i_1} \cup ... \cup C_{i_k}$ where $C_{i_1}, ..., C_{i_k}$ are disjoint, then for n_0 sufficiently large, $\Delta^{n_0}(X)_{C_j} = \sum_{h=1}^k \Delta^{n_0}(X)_{C_{i_h}}$.

For each $C \in \mathcal{C}_0$, $\Delta^n(X)^+_C$ and $\Delta^n(X)^-_C$ are positive, so $V^+(C) \geq 0, V^-(C) \geq 0$ as the weak limits of positive sequences. Define the processes on $\mathcal{A}_p$ as follows: for $A \in \mathcal{A}_p$

$$V^+_A = \sum_{\substack{C \in \mathcal{C}^\ell(\mathcal{A}_p) \\ C \subseteq A}} V^+(C), \quad V^-_A = \sum_{\substack{C \in \mathcal{C}^\ell(\mathcal{A}_p) \\ C \subseteq A}} V^-(C).$$

By additivity of V on $\mathcal{C}_0$, V is well-defined on $\mathcal{A}_0$. A straightforward calculation shows that $V_C^+ = V^+(C)$ and $V_C^- = V^-(C)$ for all $C \in \mathcal{C}_0$. It follows that V^+ and V^- are $\mathcal{A}_0$-indexed increasing processes. They are adapted since $\Delta^n(X)_A$ is $\mathcal{F}_A$-measurable, $\forall A \in \mathcal{A}$. We set $V_A = V_A^+ - V_A^-$ for each $A \in \mathcal{A}_0$. Then V is the difference of two increasing processes on $\mathcal{A}_0$.

Henceforth, we will write $a(n) = n^{(n)}$. As we have seen, the process $M = \{M_A = X_A - V_A, A \in \mathcal{A}_0\}$ is adapted, and we claim that it is a weak martingale on $\mathcal{A}_0$. Indeed, for $C \in \mathcal{C}_0$, let ξ be a $\mathcal{G}_C$-measurable bounded random variable. It follows by weak L^1-convergence that,

$$E(\xi V_C) = \lim_{n \to \infty} E[\xi \sum_{i=1}^{k(a(n))} E(X_{C \cap C_i^{a(n)}} \mid \mathcal{G}_{C \cap C_i^{a(n)}})]$$
$$= E(\xi X_C).$$

Then for any $C \in \mathcal{C}_0$, $E(M_C \mid \mathcal{G}_C) = 0$. Finally, for any $A \in \mathcal{A}$ we define

$$V_A^+ = \lim_{n \to \infty} (V_{g_n(A)})^+, V_A^- = \lim_{n \to \infty} (V_{g_n(A)})^- \text{ and } V_A = V_A^+ - V_A^-.$$

The process $\{V_A, A \in \mathcal{A}\}$ has bounded variation as a limit of the difference of two increasing processes, and is monotone outer-continuous a.s. and in L^1. Moreover, $\{M_A = X_A - V_A, A \in \mathcal{A}\}$ is a weak quasimartingale. Indeed, for $C = A \setminus \cup_{i=1}^k B_i \in \mathcal{C}$ define $C_m = g_m(A) \setminus \cup_{i=1}^k g_m(B_i)$. Then C_m is a finite union of sets from $\mathcal{C}^\ell(\mathcal{A}_m)$ and it follows that $X_{C_m} \to X_C$ a.s. and in L^1 as $m \to \infty$ by (L^1-) monotone outer-continuity of X. Since $\mathcal{G}_C$ is decreasing in $\mathcal{C}$, we may assume without loss of generality that $C_m \in \mathcal{C}^\ell(\mathcal{A}_m)$. Then

$$E(X_C \mid \mathcal{G}_C) = \lim_{m \to \infty} E(X_{C_m} \mid \mathcal{G}_C) \text{ a.s. and in } L^1$$
$$\text{(by the } (L^1-)\text{monotone outer-continuity of } X)$$
$$= \lim_{m \to \infty} E[E(X_{C_m} \mid \mathcal{G}_{C_m}) \mid \mathcal{G}_C]$$
$$\text{(since } \mathcal{G}_C \subseteq \mathcal{G}_{C_m})$$
$$= \lim_{m \to \infty} E(V_{C_m} \mid \mathcal{G}_C) = E(V_C \mid \mathcal{G}_C),$$

which finishes the proof of the theorem. $\square$

Now we extend a classic result of Rao (1969), but we no longer assume that $T \in \mathcal{A}$.

QUASIMARTINGALES 127

Definition 6.3.11 *If for a quasimartingale $X = \{X_A, A \in \mathcal{A}\}$ we have $\lim_{A \uparrow T} E(|X_A|) = 0$, we will say that X is a quasipotential.*

Theorem 6.3.12 *(Riesz decomposition of a quasimartingale) Suppose that SHAPE and Assumption 3.2.2 hold and that $\sup_{A \in \mathcal{A}} \tilde{Q}(X, A) < \infty$. Then every quasimartingale X can be written as a sum of a martingale Y and a quasipotential Z, and this decomposition is unique in the sense of stochastic equivalence, i.e., if there are two decompositions $X = Y^1 + Z^1 = Y^2 + Z^2$, then $\forall A \in \mathcal{A}$, $P(Y_A^1 = Y_A^2) = 1$, and $P(Z_A^1 = Z_A^2) = 1$.*

Proof. Uniqueness. If $X_A = Y_A^1 + Z_A^1 = Y_A^2 + Z_A^2$ are two Riesz decompositions of X, we must have $E(|Y_A^1 - Y_A^2|) \to 0$ as $A \uparrow T$. But since $|Y^1 - Y^2|$ is a submartingale, $E(|Y_A^1 - Y_A^2|)$ is a nondecreasing function of A. It follows that $E(|Y_A^1 - Y_A^2|) \equiv 0$, that is, $P(Z_A^1 = Z_A^2) = 1$ for every $A \in \mathcal{A}$.

Existence. Let $A_1 \subset A_2 \subset \ldots \subset A_k \subset \ldots$ be any fixed strictly increasing sequence of sets in $\mathcal{A}$ with $\cup A_n = T$. Denote

$$\Delta(i) = E(X_{A_i} - X_{A_{i-1}} \mid \mathcal{F}_{A_{i-1}}).$$

By assumption, $\sum_i E(|\Delta(i)|) \leq \sup_{A \in \mathcal{A}} \tilde{Q}(X, A) < \infty$. Let B be any fixed set in $\mathcal{A}$ and let

$$Y_B(i) = E(X_{A_i} \mid \mathcal{F}_B).$$

By Assumption 3.2.2 there exists i' such that $B \subseteq A_{i-1}$ if $i \geq i'$, and we have

$$\begin{aligned} E(\Delta(i) \mid \mathcal{F}_B) &= E(E(X_{A_i} \mid \mathcal{F}_{A_{i-1}}) - X_{A_{i-1}} \mid \mathcal{F}_B) \\ &= Y_B(i) - Y_B(i-1). \end{aligned}$$

Therefore, if $i \geq i'$

$$E(|Y_B(i) - Y_B(i-1)|) \leq E(|\Delta(i)|) \quad \text{(since } B \subseteq A_{i-1}\text{)},$$

which implies that

$$\sum_i E(|Y_B(i) - Y_B(i-1)|) < \infty.$$

It follows that

$$Y_B = \lim_{i \to \infty} Y_B(i)$$

exists almost surely and in L^1. The L^1-convergence of $Y_B(i)$ implies for $B \supseteq A$

$$E(Y_B \mid \mathcal{F}_A) = \lim_{i \to \infty} E(Y_B(i) \mid \mathcal{F}_A)$$

$$= \lim_{i\to\infty} E(E(X_{A_i} \mid \mathcal{F}_B) \mid \mathcal{F}_A)$$
$$= \lim_{i\to\infty} E(X_{A_i} \mid \mathcal{F}_A) = \lim_{i\to\infty} Y_A(i) = Y_A,$$

that is, Y is a martingale (additivity follows from SHAPE).

Now we will show that
$$E(|X_{A_i} - Y_{A_i}|) \to 0 \quad \text{as} \quad A_i \uparrow T.$$

Given $\epsilon > 0$ we can choose i_0 large enough that
$$\sum_{j \geq i_0} E(|\Delta(j)|) < \frac{1}{2}\epsilon.$$

Let $k \geq i_0$. Since Y_{A_k} is the limit in L^1 of $Y_{A_k}(m)$, we can choose $m - 1$ (depending on k and greater than k) with
$$E(|Y_{A_k} - Y_{A_k}(m-1)|) < \frac{1}{2}\epsilon.$$

We then have
$$E(|X_{A_k} - Y_{A_k}|) \leq E(|X_{A_k} - Y_{A_k}(m-1)|) + \frac{1}{2}\epsilon. \qquad (6.5)$$

Also, for $j - 1 \geq k$,
$$\begin{aligned} E(\Delta(j) \mid \mathcal{F}_{A_k}) &= E(E(X_{A_j} - X_{A_{j-1}} \mid \mathcal{F}_{A_{j-1}}) \mid \mathcal{F}_{A_k}) \\ &= E(X_{A_j} \mid \mathcal{F}_{A_k}) - E(X_{A_{j-1}} \mid \mathcal{F}_{A_k}) \\ &= Y_{A_k}(j) - Y_{A_k}(j-1), \end{aligned}$$

so that
$$E(|\Delta(j)|) \geq E(|Y_{A_k}(j) - Y_{A_k}(j-1)|).$$

Hence
$$\begin{aligned} E(|X_{A_k} - Y_{A_k}(m-1)|) &\leq \sum_{k<j<m} E(|Y_{A_k}(j) - Y_{A_k}(j-1)|) \\ &\leq \sum_{k<j<m} E(|\Delta(j)|) \\ &\leq \sum_{j \geq i_0} E(|\Delta(j)|) \\ &\leq \frac{1}{2}\epsilon \end{aligned}$$

by the choice of i_0. Applying the last inequality to (6.5), we have that
$$\lim_{i\to\infty} E(|X_{A_i} - Y_{A_i}|) = 0. \qquad (6.6)$$

By this we have shown the following: to any sequence (A_i) strictly increasing to T there corresponds a martingale Y satisfying (6.6). To complete the proof, we must show that (6.6) holds for any sequence $B_i \uparrow T$.

The proof will be given in three steps.

(i) We will first show that if Y is the martingale associated with the family (A_i) such that $\lim_{i \to \infty} E|X_{A_i} - Y_{A_i}| = 0$, then the same limit will hold along any sequence (B_i) such that $B_i \subseteq A_i$ for each i and $B_i \uparrow T$.

So, one has to show: $\lim_{i \to \infty} E|X_{B_i} - Y_{B_i}| = 0$. But

$$\begin{aligned}
E|X_{B_i} - Y_{B_i}| &= E|E(X_{B_i} - Y_{B_i} \mid \mathcal{F}_{B_i})| \\
&\leq E|E(X_{A_i} - X_{B_i} \mid \mathcal{F}_{B_i})| \\
&\quad + E|E(X_{A_i} - Y_{A_i} \mid \mathcal{F}_{B_i})| \\
&\leq E|E(X_{A_i} - X_{B_i} \mid \mathcal{F}_{B_i})| + E|X_{A_i} - Y_{A_i}|,
\end{aligned}$$

where the last term of the right hand side tends to zero by assumption.

It remains to show that $E|E(X_{A_i} - X_{B_i} \mid \mathcal{F}_{B_i})| \to 0$ as $i \to \infty$. Suppose on the contrary that there exists $\epsilon > 0$ such that $E|E(X_{A_i} - X_{B_i} \mid \mathcal{F}_{B_i})| > \epsilon$ infinitely often. Then by Assumption 3.2.2 there exists a subsequence (i_j) such that $E|E(X_{A_{i_j}} - X_{B_{i_j}} \mid \mathcal{F}_{B_{i_j}})| > \epsilon \ \forall j$ and for $j > 1$, $A_{i_{j-1}} \subseteq B_{i_j}$. Letting $D_{2j-1} = B_{i_j}$ and $D_{2j} = A_{i_j}$ for $j \geq 1$, we have an increasing sequence (D_j) in $\mathcal{A}$ such that

$$\sum_j E|E(X_{D_{j+1}} - X_{D_j} \mid \mathcal{F}_{D_j})|$$
$$\geq \sum_j E|E(X_{A_{i_j}} - X_{B_{i_j}} \mid \mathcal{F}_{B_{i_j}})| = \infty.$$

Thus, by contradiction $E|E(X_{A_i} - X_{B_i} \mid \mathcal{F}_{B_i})| \to 0$ as $i \to \infty$.

It follows that $\lim_i E|X_{B_i} - Y_{B_i}| = 0$ for each sequence (B_i) such that $B_i \uparrow T$ and $B_i \subseteq A_i \ \forall i$.

(ii) Now, let us show this property for the opposite inclusion: i.e., $A_i \subseteq B_i \ \forall i$. Let Y be the martingale associated with (A_i) such that $\lim_i E|X_{A_i} - Y_{A_i}| = 0$. Let R be the martingale associated with (B_i) such that $\lim_i E|X_{B_i} - R_{B_i}| = 0$. Now,

$$E|Y_{A_i} - R_{A_i}| \leq E|X_{A_i} - Y_{A_i}| + E|X_{A_i} - R_{A_i}|,$$

where the first term on the right tends to zero by assumption and the second term on the right tends to zero by the previous step.

Since $(Y - R)$ is a martingale, $E|Y_A - R_A|$ is a non-decreasing function of A and by Assumption 3.2.2, $A \subseteq A_i$ for all i sufficiently large. Therefore, since $E|Y_{A_i} - R_{A_i}| \to 0$, this implies that $P(Y_A = R_A) = 1$, $\forall A \in \mathcal{A}$.

(iii) We now show that $E|X_{B_n} - Y_{B_n}| \to 0$ as $B_n \uparrow T$, for any sequence (B_n). One can easily show that the sequence $(A_i \cap B_i)$ is strictly increasing to T and lies in $\mathcal{A}$. But $A_i \cap B_i \subseteq A_i$, $\forall i$, which implies that the limit holds for $(A_i \cap B_i)$, i.e. $\lim_i E|X_{A_i \cap B_i} - Y_{A_i \cap B_i}| = 0$ (step 1). Furthermore, $B_i \supseteq A_i \cap B_i$, $\forall i$. It follows that $\lim_{B_i \uparrow T} E|X_{B_i} - Y_{B_i}| = 0$ (step 2). To complete the proof, for each $A \in \mathcal{A}$ let $Z_A = X_A - Y_A$. $\square$

PART II

Weak Convergence

CHAPTER 7

Weak Convergence of Set-Indexed Processes

We now turn from the general theory of set-indexed processes to the question of weak convergence of a sequence of set-indexed processes. To do so, we shall define $\mathcal{D}(\mathcal{A})$, a set-indexed analogue of the Skorokhod function space $D([0,\infty))$. This work is motivated by that of Bass and Pyke (1985), who worked with a class of processes indexed by subsets of the d-dimensional unit cube. As will be observed, the topology appropriate for the set-indexed framework corresponds to the Skorokhod J_2 topology, and not to the better-known J_1 topology. Sufficient conditions for tightness will be given, and we shall see that tightness combined with convergence of finite dimensional distributions is equivalent to weak convergence on the function space.

The class of processes whose sample paths lie in $\mathcal{D}(\mathcal{A})$ is very large, and includes most purely atomic processes. However, as will be seen, if the class $\mathcal{A}$ is too large (in the sense of metric entropy), then it may not include the Brownian motion. As this precludes functional central limit theorems over such classes $\mathcal{A}$, we shall introduce a second mode of convergence for set-indexed processes: *semi-functional convergence*, which implies functional convergence over flows. It will be seen that semi-functional convergence is ideally suited to general set-indexed processes, and while it imposes no restrictions on the metric entropy of the class $\mathcal{A}$, it does imply convergence of finite dimensional distributions.

A metric is now required: therefore, throughout Part II *we shall assume that (T,d) is a compact complete separable metric space, and that T is in $\mathcal{A}(u)$*. Extension of all results to σ-compact metric spaces is straightforward, and will be discussed briefly. As well, we shall assume (cf. Assumption 1.3.2) that the sub-semilattices $\mathcal{A}_n$ and the functions g_n may be chosen so that there exists a sequence

(ϵ_n), $\epsilon_n \downarrow 0$ such that for each $n \in \mathbf{N}$,
$$d_H(A, g_n(A)) \leq \epsilon_n, \ \forall A \in \mathcal{A}.$$

The material in this chapter is based primarily on Slonowsky (1998).

7.1 The Function Space $\mathcal{D}(\mathcal{A})$

Recall the definition of the Hausdorff metric d_H defined on $\mathcal{K}\setminus\emptyset$, the nonempty compact subsets of T (cf. Section 1.3): for nonempty sets $A, B \in \mathcal{K}$,
$$d_H(A, B) = \inf\{\epsilon > 0 : A \subseteq B^\epsilon \text{ and } B \subseteq A^\epsilon\}.$$

Also, we recall from Lemma 1.3.3 that the indexing collection $\mathcal{A}$ is d_H-closed when T is compact and Assumption 1.3.2 is satisfied.

For any function $x : \mathcal{A} \to \mathbf{R}$, define
$$\| x \|_\mathcal{A} = \sup_{A \in \mathcal{A}} | x(A) |$$
and let
$$B(\mathcal{A}) = \{x : \mathcal{A} \to \mathbf{R}; \| x \|_\mathcal{A} < \infty\}.$$

We note that $(B(\mathcal{A}), \| \cdot \|_\mathcal{A})$ is complete. Let $C(\mathcal{A})$ denote the class of functions in $B(\mathcal{A})$ which are d_H-continuous on $\mathcal{A}$.

Definition 7.1.1 *For any $A, (A_n)_n$ in $\mathcal{A}$, we write*

(a) $A_n \searrow A$ *if* $A \subseteq A_n$ $\forall n$ *and* $A_n \xrightarrow{d_H} A$ *and*

(b) $A_n \nearrow A$ *if* $A_n \subseteq A^\circ$ $\forall n$ *and* $A_n \xrightarrow{d_H} A$

Given $x : \mathcal{A} \to \mathbf{R}$, we say

(c) x *is* outer-continuous *at A if $A_n \searrow A$ implies $x(A_n) \to x(A)$ and*

(d) x *has* inner limits *at A if $A_n \nearrow A$ implies $(x(A_n))_n$ converges.*

If x is outer-continuous (has inner limits) at every $A \in \mathcal{A}$, then we simply say that x is outer-continuous (respectively, has inner limits). We denote by $\mathcal{D}(\mathcal{A})$ the class of functions in $B(\mathcal{A})$ which are outer-continuous with inner limits.

Comments 7.1.2 1. For $T = [0,1]^d$ and $\mathcal{A}$ any indexing collection, this is exactly the definition of Bass and Pyke (1985).
2. We note the distinction between functions which are *monotone*

outer-continuous and functions which are outer-continuous as defined above. Clearly, outer-continuous functions are trivially monotone outer-continuous on $\mathcal{A}$. In Part I, T was usually assumed to be a general topological space, and so the idea of a sequence of sets 'converging' only made sense if the sequence was monotone. However, when T is a metric space, we can introduce the stronger mode of convergence on non-monotone sequences.

3. Monotone inner-continuity on $\mathcal{A}$ was defined in Definition 5.1.4. If $A_n \nearrow A$ implies $(x(A_n))_n \to x(A)$, then we say that x is inner-continuous. However, note that inner-continuity does not necessarily imply monotone inner-continuity, since $A_n \nearrow A$ implies that $A_n \subseteq A^\circ$. Monotone inner-continuity required that the sequence (A_n) be monotone increasing with $A_n \subseteq A$, but it was not assumed that $A_n \subseteq A^\circ$. However, it is straightforward to show that under an assumption of separability from within, that any inner-continuous function x which is 0 on all sets with empty interior is monotone inner-continuous in the sense of Definition 5.1.4. In particular, we need only assume that if $A \in \mathcal{A}$ has non-empty interior, then for every $\epsilon > 0$, there exists $A_\epsilon \in \mathcal{A}$, such that $A_\epsilon \subseteq A^\circ$ and $d_H(A_\epsilon, A) < \epsilon$. Then if (A_n) is an increasing sequence with $\overline{\cup_n A_n} = A$ and (without loss of generality) $A_n^\circ \neq \emptyset$, $\forall n$, there exists a sequence $(A_n^{'})$ such that $A_n^{'} \subseteq A_n^\circ$, $d_H(A_n^{'}, A) < 1/n$, and $\mid x(A_n) - x(A_n^{'}) \mid < 1/n$. Thus, $A_n^{'} \nearrow A$, and so $x(A_n^{'}) \to x(A)$. But it is an immediate consequence that $x(A_n) \to x(A)$, proving monotone inner-continuity.

4. Note that inner- and outer-continuity of a function x do not necessarily imply that x is d_H-continuous on $\mathcal{A}$: i.e. that if $A, A_1, A_2, \ldots \in \mathcal{A}$ and if $d_H(A_n, A) \to 0$, then $x(A_n) \to x(A)$. Example 7.1.3 #1 below provides a counterexample.

Examples 7.1.3 1. For $T = [0, 1]^d$ and $\mathcal{A} = \{[0, t] : t \in T\}$, the class $\mathcal{D}(\mathcal{A})$ is exactly $D([0, 1])$ if $d = 1$ (cf. Skorokhod (1956), Billingsley (1968)), but is strictly larger than $D([0, 1]^d)$ for $d > 1$ (cf. Neuhaus (1971), Straf (1972)). To see this for $d = 2$, consider the indicator function I of the set $\{(t_1, t_2) : t_1 + t_2 \geq k\}$, $0 < k \leq 1$. Any function $x \in D([0, 1]^2)$ must have limits at each point $t = (t_1, t_2)$ from all four quadrants, so clearly $I \notin D([0, 1]^2)$, but $I \in \mathcal{D}(\mathcal{A})$.

2. Trivially, $C(\mathcal{A}) \subset \mathcal{D}(\mathcal{A})$.

3. If $T = \mathbf{R}_+^d$ or $\mathbf{R}^d$ and $\mathcal{A}$ is a Vapnik-Červonenkis class in $\sigma(\mathcal{A})$, then the Brownian motion with variance measure Λ is in

$\mathcal{D}(\mathcal{A})$ provided that Λ is. *However, as noted in Section 3.4, this is no longer the case if $\mathcal{A}$ is the class of lower sets.*

4. We now consider the class of *purely atomic* functions which include the sample paths of point processes.

Definition 7.1.4 *A set-function $x : \mathcal{A} \to \mathbf{R}$ is purely atomic if, for some $n \in \mathbf{N}$, $\exists\, t_1, \cdots, t_n \in T$ and $a_1, \cdots, a_n \in \mathbf{R}$ such that*

$$x(A) = \sum_{j\,:\,t_j \in A} a_j, \quad (\forall A \in \mathcal{A}).$$

The t_j are called the locations *of atoms and the a_j are called the* masses *of atoms. The variation of x is the number, $\nu(x) = \sum_{i=1}^{n} |a_i|$.*

It is tempting to conclude that all purely atomic functions are in $\mathcal{D}(\mathcal{A})$, but this is not the case. Consider $T = [0,1]^2$ and $\mathcal{A} = \{[0,t] : t \in T\}$. Let x be the function with a single atom of mass 1 at $(1, 1/2)$. Then x does not have an inner limit at any point $(1, v)$ if $1/2 < v \leq 1$. To obtain a subclass of purely atomic functions which are in $\mathcal{D}(\mathcal{A})$, we introduce the concept of *proper sets* which was first defined in Ivanoff and Merzbach (1999).

Definition 7.1.5 (a) *Given $C \subseteq T$ and $\epsilon > 0$, define $C^{-\epsilon} := \cap_{B \in \mathcal{A},\, C \subseteq B^\epsilon} B$.*
(b) *$A \in \mathcal{A}$ is* proper *in $\mathcal{A}$ provided $\exists\, \epsilon_0 > 0$ such that $(A^\epsilon)^{-\epsilon} = A$ $\forall\, 0 < \epsilon < \epsilon_0$.*

We note that by definition, $(A^\epsilon)^{-\epsilon} \subseteq A$ for any $A \in \mathcal{A}$. Also, if $A_n \nearrow A$, then for any $\epsilon > 0$, $A^{-\epsilon} \subseteq A_n$ for all n sufficiently large. It is not true that every set in $\mathcal{A}$ is proper. For example, if $T = [0,1]^2$ and $\mathcal{A} = \{[0,t] : t \in T\}$, then $[0,t]$ is not proper if and only if $t = (x_1, x_2)$ where $x_1 = 1$ or $x_2 = 1$.

Now, for $t \in T$, define

$$A_t = \cap_{\substack{A \in \mathcal{A} \\ t \in A}} A. \qquad (7.1)$$

We define $\mathcal{P}(\mathcal{A})$ to be the class of purely atomic functions x such that A_t is proper for each $t \in T$ which is the location of an atom of x. It was proved by Slonowsky (1998) that $\mathcal{P}(\mathcal{A}) \subset \mathcal{D}(\mathcal{A})$ and that every $x \in \mathcal{P}(\mathcal{A})$ is d_H-continuous at $\emptyset'$ and T:

Theorem 7.1.6 *(Slonowsky (1998))*
Let $x : \mathcal{A} \to \mathbf{R}$ be a purely atomic set-function with atoms located at $t_1, \cdots, t_n \in T$. If A_{t_i} is proper $\forall\, 1 \leq i \leq n$, then $x \in \mathcal{D}(\mathcal{A})$.

Proof. For any $S \subseteq T$, define $P_S = \{j : t_j \in S\}$. To establish

THE FUNCTION SPACE $\mathcal{D}(\mathcal{A})$ 137

outer-continuity, fix $A \in \mathcal{A}$ and take $A_n \searrow A$ in $\mathcal{A}$. Since A is d-closed in T, $\exists \epsilon_0 > 0$ such that

$$d(A, t_j) \geq \epsilon_0, \ (\forall j \notin P_A).$$

By the definition of P_S, this implies $P_A = P_{A^{\epsilon_0}}$. Since $A_n \to_{d_H} A$, $\exists K \in \mathbf{N}$ such that $A_n \subseteq A^{\epsilon_0} \ \forall n \geq K$ and thus,

$$P_{A_n} \subseteq P_{A^{\epsilon_0}} = P_A, \ (\forall n \geq K).$$

But $A \subseteq A_n \ \forall n$ implies $P_A \subseteq P_{A_n} \ \forall n$. Therefore, $P_{A_n} = P_A$ $\forall n \geq K$, and so $x(A_n) \to x(A)$, establishing outer-continuity of x at A. (The above argument works for any purely atomic x, regardless of the locations of its atoms.)

To establish that x has inner limits on $\mathcal{A}$, first consider the case in which x has a single atom of mass 1 at some $t \in T$ such that A_t is proper. Defining the *set-interval*,

$$[A_t, T] = \{A \in \mathcal{A} : A_t \subseteq A \subseteq T\},$$

it is clear from the definition of A_t that $x = \mathbf{1}_{[A_t, T]}$. Furthermore, it is shown in Lemma 2.5 of Ivanoff and Merzbach (1999) that if A_t is a proper set, then $x = \mathbf{1}_{[A_t, T]}$ has inner limits on $\mathcal{A}$. To be precise, for $A, (A_n)_n$ in $\mathcal{A}$, if $A_n \nearrow A$, then if $A_t \not\subseteq A^\circ$ it follows that $A_t \not\subseteq A_n \ \forall n$ and so $x(A_n) = 0 \ \forall n$. If $A_t \subseteq A^\circ$, then since T is compact and A_t is proper, there exists $\epsilon > 0$ such that $A_t^\epsilon \subseteq A$ and $(A_t^\epsilon)^{-\epsilon} = A_t$. Thus, if $A \subseteq B^\epsilon$ for some $B \in \mathcal{A}$, then $A_t = (A_t^\epsilon)^{-\epsilon} \subseteq A^{-\epsilon} \subseteq (B^\epsilon)^{-\epsilon} \subseteq B$, and so $A_t \subseteq A^{-\epsilon}$. Therefore, $A_t \subseteq A_n$ for all n sufficiently large, and $x(A_n) \to 1$.

For a general purely atomic x with atoms $t_1, \cdots, t_n$ such that A_{t_i} is proper for every i, and respective masses $a_1, \cdots, a_n \in \mathbf{R}$, inner limits follow from the identity $x = \sum_{i=1}^n a_i x_i$ where x_i is a purely atomic set-function with a single atom of mass 1 at $t_i \in T$. □

Lemma 7.1.7 *(Slonowsky (1998))*
If $T \in \mathcal{A}$, each $x \in \mathcal{P}(\mathcal{A})$ is d_H-continuous at $\emptyset'$ and T.

Proof. Let $x \in \mathcal{P}(\mathcal{A})$ be given. Outer-continuity of x and minimality of $\emptyset'$ imply d_H-continuity at $\emptyset'$.

To establish d_H-continuity at T, consider the case in which x has a single atom of mass 1 at t such that $A_t \in \mathcal{A}$ is a proper set. Hence, there exists $\epsilon > 0$ such that $(A_t^\epsilon)^{-\epsilon} = A_t$.

Now, given $B_n \to_{d_H} T$, $\exists K \in \mathbf{N}$ such that

$$n \geq K \Rightarrow T = (B_n)^\epsilon.$$

Thus, for each $n \geq K$,
$$A_t = (A_t^\epsilon)^{-\epsilon} \subseteq T^{-\epsilon} = (B_n^\epsilon)^{-\epsilon} \subseteq B_n.$$
But $t \in A_t$ therefore, $x(B_n) = 1 = x(T)$ $\forall n \geq K$, i.e., $x(B_n) \xrightarrow{n} x(T)$, establishing that x is d_H-continuous at T.

For the general case, assume x has atoms $t_1, \cdots, t_k \in T$ such that A_{t_i} is proper for every i with respective masses, $a_1, \cdots, a_k \in \mathbf{R}$. Let $x_j \in \mathcal{P}(\mathcal{A})$ have a single atom of mass 1 at t_j $\forall 1 \leq j \leq k$. Then, given an arbitrary sequence, $B_n \to_{d_H} T$, the above case implies
$$x(B_n) = \sum_{j=1}^{k} a_j \cdot x_j(B_n) \to_n \sum_{j=1}^{k} a_j \cdot x_j(T) = x(T),$$
establishing d_H-continuity of x at T. □

We now define a metric on $\mathcal{D}(\mathcal{A})$. First, we define a metric ρ on $\mathcal{A} \times \mathbf{R}$: for $(A_1, r_1), (A_2, r_2) \in \mathcal{A} \times \mathbf{R}$,
$$\rho((A_1, r_1), (A_2, r_2)) = d_H(A_1, A_2) + |r_1 - r_2|.$$
Let $\mathcal{G}$ denote the closed and bounded subsets of $\mathcal{A} \times \mathbf{R}$, and let $d_\mathcal{G}$ denote the Hausdorff metric induced by ρ on $\mathcal{G}$. We note that $\mathcal{G}$ is complete. Now, for $x \in B(\mathcal{A})$, let $G(x)$ be the closure of $\{(A, x(A)), A \in \mathcal{A}\}$ in the metric ρ. This set is called the *incomplete closed graph* of x. It contains all points $(A, x(A))$ and (A, r) where $r = \lim x(A_n)$ for some sequence (A_n) such that $A_n \to_{d_H} A$. If $x, y \in B(\mathcal{A})$, let $D_H(x, y) = d_\mathcal{G}(G(x), G(y))$. We have
$$D_H(x, y) = \max(D_x(y), D_y(x)),$$
where
$$D_x(y) = \sup_{(A_1, r_1) \in G(x)} \inf_{(A_2, r_2) \in G(y)} \rho((A_1, r_1), (A_2, r_2)).$$

Note that D_H defines a *pseudometric* on $B(\mathcal{A})$, and a metric on $\mathcal{D}(\mathcal{A})$. Therefore, we shall identify functions $x, y \in B(\mathcal{A})$ if $D_H(x, y) = 0$.

Remarks 7.1.8 1. Other topologies may be defined on $\mathcal{D}(\mathcal{A})$ which are not equivalent to our topology. For example, if $T = [0, 1]$ and $\mathcal{A} = \{[0, t] : 0 \leq t \leq 1\}$, then our space is homeomorphic to the space $D[0, 1]$ endowed with the Skorokhod J_2 topology (Pomarede (1976), Theorem II.4.1). Although the J_1 topology is more familiar, it is not clear how to define an appropriate class of homeomorphisms on $\mathcal{A}$. (See Remark #4 in Bass and Pyke (1985)).

THE FUNCTION SPACE $\mathcal{D}(\mathcal{A})$ 139

2. Although we have assumed that T is compact, it is straightforward to show that the definition of the metric D_H can be extended to the case in which T is locally compact, following Lindvall (1973). Thus, all of the results which follow hold for locally compact metric spaces T, but we may continue to assume without loss of generality that T is compact.

It is easily seen that for $x, y \in B(\mathcal{A})$, $D_H(x, y) \leq \| x - y \|_{\mathcal{A}}$. We have the following simple result:

Proposition 7.1.9 *(i) Given $x, x_1, x_2, \ldots \in B(\mathcal{A})$, if $\| x_n - x \|_{\mathcal{A}} \to 0$, then $D_H(x_n, x) \to 0$.*
(ii) Given $x_1, x_2, \ldots \in B(\mathcal{A})$ and x continuous in the Hausdorff topology on $\mathcal{A}$, if $D_H(x_n, x) \to 0$ then $\| x_n - x \|_{\mathcal{A}} \to 0$.

In fact, uniform continuity of x is required to prove (ii) above, but this is an immediate consequence of the fact that $\mathcal{A}$ is compact in the Hausdorff topology.

Although it is easily shown that $\mathcal{D}(\mathcal{A})$ is a $\| \cdot \|_{\mathcal{A}}$-closed subspace of $B(\mathcal{A})$, $\mathcal{D}(\mathcal{A})$ is not D_H-closed. For example, if $\mathcal{D}(\mathcal{A}) = D[0, 1]$ and $x_n([0, t]) = \sin(nt)$, then as pointed out by Bass and Pyke (1985), $(x_n)_n$ is D_H-Cauchy but does not converge to a function in $\mathcal{D}(\mathcal{A})$. Therefore, although a complete characterization of the compact subsets of $\mathcal{D}(\mathcal{A})$ cannot be given, a certain class of compact sets can be described; it will be seen that these sets consist of classes of functions which can be uniformly approximated by the sum of a continuous function and a purely atomic function.

First, we define the following collection of subspaces of $\mathcal{D}(\mathcal{A})$.

Definition 7.1.10 *Given a triple,*
$$(\boldsymbol{\Gamma}, \boldsymbol{\Sigma}, \boldsymbol{\Delta}) = ((\Gamma_n)_n, (\Sigma_n)_n, (\Delta_n)_n),$$
where
$$\Gamma_n \subseteq \mathcal{P}(\mathcal{A}) \quad \text{and} \quad \Sigma_n \subseteq C(\mathcal{A}), \quad (\forall n)$$
and $(\Delta_n)_n$ is a sequence of constants such that $\Delta_n \to 0$ and $\Delta_n \geq 0$ $\forall n$, define $\Xi(\boldsymbol{\Gamma}, \boldsymbol{\Sigma}, \boldsymbol{\Delta})$ to be the collection of all $x \in B(\mathcal{A})$ such that, for each $n \in \mathbf{N}$, there exists $J_n(x) \in \Gamma_n$ and $C_n(x) \in \Sigma_n$ satisfying
$$\|x - (J_n(x) + C_n(x))\|_{\mathcal{A}} \leq \Delta_n. \tag{7.2}$$

We shall see that $\Xi(\boldsymbol{\Gamma}, \boldsymbol{\Sigma}, \boldsymbol{\Delta})$ is compact in $(\mathcal{D}(\mathcal{A}), D_H)$ whenever each Γ_n is compact in $(\mathcal{P}(\mathcal{A}), D_H)$ and each Σ_n is compact in $(C(\mathcal{A}), \| \cdot \|_{\mathcal{A}})$, provided that the boundaries of the sets in $\mathcal{A}$ are well-behaved in the following sense:

Definition 7.1.11 *An indexing collection, $\mathcal{A}$ on (T,d) is said to be consistent with boundaries (c.w.b.) if, given any $A, B \in \mathcal{A} \setminus \{T, \emptyset\}$ and any $\epsilon > 0$,*

(a) $\partial A \neq \emptyset$,

(b) $d_H(A, B) < \epsilon$ *implies* $d_H(\partial A, \partial B) < \epsilon$, *and*

(c) *given* $t \in T \setminus A$, $d(t, \partial A) \geq \epsilon$ *implies* $B_d(t, \epsilon) \cap A = \emptyset$. *($B_d(t, \epsilon)$ is the open ball of radius ϵ centred at t.)*

Remark 7.1.12 (b) and (c) give 'geometric' conditions on the sets in $\mathcal{A}$. From (b), when two sets, $A, B \in \mathcal{A}$, are close with respect to d_H, so are their boundaries. From (c), if t lies outside of A, then the points in A closest to t must lie in ∂A. Not every indexing collection is c.w.b., and counterexamples may be found in Slonowsky (1998). However, all three conditions are satisfied by all of our examples.

The reason that we must require that $\mathcal{A}$ be consistent with boundaries is to ensure that D_H-convergence in $B(\mathcal{A})$ to a limit in $\mathcal{P}(\mathcal{A})$ implies pointwise convergence on 'continuity sets'. (A nonempty set A is a continuity set of x if $x(\partial A) = 0$ (i.e., no atoms of x lie on ∂A, the boundary of A).) This result is given in Proposition 7.1.14 below. First, we need the following lemma, which is no longer true if the 'c.w.b.' assumption is dropped.

Lemma 7.1.13 *(Slonowsky (1998))*
Assume $\mathcal{A}$ is consistent with boundaries. Let $x \in \mathcal{P}(\mathcal{A})$ and let $A \in \mathcal{A}$ be a continuity set of x. Then x is is d_H-continuous (on $\mathcal{A}$) at A.

Proof. First, we observe that for any indexing collection $\mathcal{A}$ (not necessarily c.w.b.), and any purely atomic function $x : \mathcal{A} \to \mathbf{R}$, x is d_H-continuous at $A \in \mathcal{A}$, if and only if there exists $\delta > 0$ such that $x(A) = x(B)$ for every $B \in \mathcal{A}$ satisfying $d_H(A, B) < \delta$.

We recall the following notation: for $x \in \mathcal{P}(\mathcal{A})$ with atoms at $t_1, ..., t_n$, and respective weights $a_1, ..., a_n$, and $A \subseteq T$, let $P_A = \{j : t_j \in A\}$.

Observe that since $\partial T = \emptyset$, if $T \in \mathcal{A}$, T is automatically a continuity set for any $x \in \mathcal{P}(\mathcal{A})$, and Lemma 7.1.7 implies that x is d_H-continuous at T.

If $A \in \mathcal{A} \setminus \{T\}$ is a continuity set for $x \in \mathcal{P}(\mathcal{A})$, then, since $P_{\partial A} = \emptyset$ and $\partial A \neq \emptyset$ is d-closed in T,

$$\delta = (1/2) \cdot \min_{1 \leq j \leq n} d(t_j, \partial A) > 0. \qquad (7.3)$$

Now take $B \in \mathcal{A}$ such that $d_H(A, B) < \delta$. To show $x(A) = x(B)$,

THE FUNCTION SPACE $\mathcal{D}(\mathcal{A})$ 141

it is clearly sufficient to show $P_A = P_B$. First, for the inclusion $P_B \subseteq P_A$, take j such that $t_j \notin A$ (if such a j exists). Since $d_H(A, B) < \delta < d(t_j, \partial A)$, (c) of Definition 7.1.11 ensures that $t_j \notin B$.

To establish the opposite inclusion, take j such that $t_j \notin B$ (if such a j exists). There are two cases. First, if $d(t_j, \partial B) \geq \delta$, then, by the same argument as above, $t_j \notin A$, i.e., $j \notin P_A$.

On the other hand, if $d(t_j, \partial B) < \delta$, then $\exists \tilde{b} \in \partial B$ such that $d(t_j, \tilde{b}) < \delta$. Since $\mathcal{A}$ is c.w.b.,

$$d_H(A, B) < \delta \implies d_H(\partial A, \partial B) < \delta$$

and thus, $\partial B \subseteq (\partial A)^\delta$. This implies $\exists \tilde{a} \in \partial A$ such that $d(\tilde{b}, \tilde{a}) \leq \delta$ which, by (7.3), implies

$$\begin{aligned} d(t_j, \partial A) &= \inf_{a \in \partial A} d(t_j, a) \leq d(t_j, \tilde{a}) \\ &\leq d(t_j, \tilde{b}) + d(\tilde{b}, \tilde{a}) < 2 \cdot \delta \leq d(t_j, \partial A), \end{aligned}$$

resulting in a contradiction. Therefore, if $d_H(A, B) < \delta$, it is impossible for t_j to satisfy $d(t_j, \partial B) < \delta$. This establishes the inclusion, $P_A \subseteq P_B$, completing the proof of the lemma. □

Proposition 7.1.14 *(Slonowsky (1998))*
Assume that $\mathcal{A}$ is consistent with boundaries and let (x_n) be any sequence of functions in $B(\mathcal{A})$. If $x_n \to_{D_{II}} x$ for some $x \in \mathcal{P}(\mathcal{A})$, then for every continuity set $A \in \mathcal{A}$ of x,

$$x_n(A) \to x(A).$$

Proof. Assume that there exists a continuity set $A \in \mathcal{A}$ of x, such that $x_n(A) \not\to x(A)$. Then, $\exists \alpha > 0$ and a subsequence, $(x_{k_n})_n$ such that

$$|x_{k_n}(A) - x(A)| \geq \alpha, \quad (\forall n). \tag{7.4}$$

Via Lemma 7.1.13 we can assume, without a loss of generality, that α is small enough so that $d_H(A, B) < \alpha$ implies $x(A) = x(B)$. We shall show that for every $n \in \mathbf{N}$,

$$G(x_{k_n}) \not\subseteq [G(x)]^{\alpha/4}, \tag{7.5}$$

which implies that $x_n \not\to_{D_{II}} x$. Thus, the proposition follows by contradiction.

It remains to prove that (7.4) implies (7.5). Assume $\exists n \in \mathbf{N}$ such that $G(x_{k_n}) \subseteq [G(x)]^{\alpha/4}$. Then, by the definition of $G(x)$, $\exists B \in \mathcal{A}$ such that $\rho((A, x_{k_n}(A)), (B, x(B))) \leq \alpha/2$. Therefore,

$$d_H(A, B) \leq \alpha/2 \text{ and } |x_{k_n}(A) - x(B)| \leq \alpha/2. \tag{7.6}$$

But, as mentioned above, α has been chosen small enough to ensure that $d_H(A,B) < \alpha$ implies $x(A) = x(B)$. Therefore, by (7.4) and (7.6),

$$\alpha \leq |x_{k_n}(A) - x(A)| = |x_{k_n}(A) - x(B)| \leq \alpha/2,$$

which gives us a contradiction and thus establishes (7.5). $\square$

The proof of the following theorem is very close to that of Theorem 3.4 of Bass and Pyke (1985), and so only a sketch will be given here. Note that here it is only assumed Γ_n is compact, whereas Bass and Pyke require Γ_n to be of a particular form which ensures that D_H-convergence in $B(\mathcal{A})$ to a limit in $\mathcal{P}(\mathcal{A})$ implies pointwise convergence on 'continuity sets'. Proposition 7.1.14 permits the more general formulation below.

Theorem 7.1.15 *(Slonowsky (1998))*
Assume that $\mathcal{A}$ is consistent with boundaries. Given a triple (Γ, Σ, Δ) as in Definition 7.1.10, if
1. *Γ_n is compact in $(\mathcal{P}(\mathcal{A}), D_H)$ $\forall n$ and*
2. *Σ_n is compact in $(C(\mathcal{A}), \|\cdot\|_{\mathcal{A}})$ $\forall n$,*

then $\Xi(\Gamma, \Sigma, \Delta)$ is compact in $(\mathcal{D}(\mathcal{A}), D_H)$.

Proof. (Sketch):
In what follows, we let $\Xi = \Xi(\Gamma, \Sigma, \Delta)$. Since G is an isometric embedding, it is enough to show that $G(\Xi) := \{G(x) : x \in \Xi\}$ is a compact subset of $\mathcal{G}$. However, (cf. Slonowsky (1998), Bass and Pyke (1985)), this will follow if it can be shown that $G(\Xi)$ is d_G-closed and that $\sup_{x \in \Xi} \|x\|_{\mathcal{A}} < \infty$.

It is an easy consequence of the fact that Γ_n and Σ_n are compact that $\sup_{x \in \Xi} \|x\|_{\mathcal{A}} < \infty$, and so it only remains to prove that $G(\Xi)$ is d_G-closed. Thus, take a sequence (x_n) in Ξ such that $G(x_n) \to G_0$ for some $G_0 \in \mathcal{G}$. It must be shown that there exists a function $y \in \Xi$ such that $G(y) = G_0$.

First, for a fixed $n \in \mathbf{N}$, we may select sequences $(J_m(x_n))_m$ in $\mathcal{P}(\mathcal{A})$ and $(C_m(x_n))_m$ in $C(\mathcal{A})$ according to Definition 7.1.10, such that $J_m(x_n) \in \Gamma_m$ and $C_m(x_n) \in \Sigma_m$ for every m. Now, fix m. Since Γ_m and Σ_m are assumed to be compact, there exists a subsequence (x_{k_n}) and elements $j_m \in \Gamma_m$ and $c_m \in \Sigma_m$ such that as $n \to \infty$,

$$J_m(x_{k_n}) \to_{D_H} j_m \text{ and } \|C_m(x_{k_n}) - c_m\|_{\mathcal{A}} \to 0. \qquad (7.7)$$

Since m is arbitrary, in this way we may construct sequences (j_m)

THE FUNCTION SPACE $\mathcal{D}(\mathcal{A})$

and (c_m) such that for each m, $j_m \in \Gamma_m$ and $c_m \in \Sigma_m$ and (7.7) is satisfied. By a diagonalization argument, the same subsequence may be chosen for each m. The remainder of the proof is carried out in three steps:

1. There exists a function $y \in B(\mathcal{A})$ such that $\| y-(j_m+c_m) \|_{\mathcal{A}} \to 0$ as $m \to \infty$.

2. The function y lies in Ξ.

3. $G(y) = G_0$.

Proof of Step 1: Since $(B(\mathcal{A}), \| \cdot \|_{\mathcal{A}})$ is complete, it is enough to show that $(j_m + c_m)$ is $\| \cdot \|_{\mathcal{A}}$-Cauchy. This is shown by a long but elementary proof by contradiction. The proof uses the fact that by (7.7), if m is fixed, for any $A \in \mathcal{A}$, $C_m(x_{k_n})(A) \to_n c_m(A)$ and for any $A \in \mathcal{A}$ and $\delta > 0$, since $\mathcal{A}$ is c.w.b. there exists a continuity set B of j_m such that $d_H(A, B) < \delta$, $j_m(A) = j_m(B)$, and $J_m(x_{k_n})(B) \to_n j_m(B)$. Thus, if $(j_m + c_m)$ is not $\| \cdot \|_{\mathcal{A}}$-Cauchy, there exists $\epsilon > 0$ such that for every M, there exist $m, h > M$ and $A \in \mathcal{A}$ such that

$$| (j_m(A) + c_m(A)) - (j_h(A) + c_h(A)) | > \epsilon. \tag{7.8}$$

However, we can choose a continuity set B of both j_m and j_h close enough to A such that $j_m(A) = j_m(B)$, $j_h(A) = j_h(B)$, and for all n sufficiently large

$$\begin{aligned}
& | (j_m(A) + c_m(A)) - (j_h(A) + c_h(A)) | \\
\leq\ & | (j_m(B) + c_m(B)) - (j_h(B) + c_h(B)) | + \\
& | c_m(A) - c_m(B) | + | c_h(A) - c_h(B) | \\
\leq\ & | (J_m(x_{k_n})(B) + C_m(x_{k_n})(B)) - \\
& (J_h(x_{k_n})(B) + C_h(x_{k_n})(B)) | + \\
& | j_m(B) - J_m(x_{k_n})(B) | + | j_h(B) - J_h(x_{k_n})(B) | + \\
& | c_m(B) - C_m(x_{k_n})(B) | + | c_h(B) - C_h(x_{k_n})(B) | + \\
& | c_m(A) - c_m(B) | + | c_h(A) - c_h(B) | \\
\leq\ & | x_{k_n}(B) - (J_m(x_{k_n})(B) + C_m(x_{k_n})(B)) | + \\
& | x_{k_n}(B) - (J_h(x_{k_n})(B) + C_h(x_{k_n})(B)) | + \\
& | j_m(B) - J_m(x_{k_n})(B) | + | j_h(B) - J_h(x_{k_n})(B) | + \\
& | c_m(B) - C_m(x_{k_n})(B) | + | c_h(B) - C_h(x_{k_n})(B) | + \\
& | c_m(A) - c_m(B) | + | c_h(A) - c_h(B) | \\
<\ & \epsilon,
\end{aligned}$$

which contradicts (7.8).

Proof of Step 2: By the definition of the sequences (j_m) and (c_m), it is enough to prove that $\| y - (j_m + c_m) \|_{\mathcal{A}} \leq \Delta_m$, for every m. A straightforward argument gives the following:

$$\begin{aligned}
\| y - (j_m + c_m) \|_{\mathcal{A}} & \\
\leq \ & \limsup_h \| (j_h + c_h) - (j_m + c_m) \|_{\mathcal{A}} \\
\leq \ & \limsup_h \limsup_n \| (J_h(x_{k_n}) + C_h(x_{k_n})) - \\
& (J_m(x_{k_n}) + C_m(x_{k_n})) \|_{\mathcal{A}} \\
\leq \ & \limsup_h (\limsup_n (\Delta_h + \Delta_m)) \\
= \ & \Delta_m.
\end{aligned}$$

Proof of Step 3: It suffices to show that $d_G(G(y), G_0) < \epsilon$, for every $\epsilon > 0$. We observe that for any m, n,

$$\begin{aligned}
d_G(G(y), G_0) \leq \ & d_G(G(y), G(j_m + c_m)) + \\
& d_G(G(j_m + c_m), G(J_m(x_{k_n}) + C_m(x_{k_n}))) + \\
& d_G(G(J_m(x_{k_n}) + C_m(x_{k_n})), G(x_{k_n})) + \\
& d_G(G(x_{k_n}), G_0).
\end{aligned}$$

Now, given $\epsilon > 0$, we may choose m and then $n = n(m)$ sufficiently large that each of the terms on the right hand side of the above equation is less than $\epsilon/4$. This completes the proof. □

7.2 Weak Convergence on $\mathcal{D}(\mathcal{A})$

In this section we consider criteria for weak convergence of a sequence of processes (X_n) in $\mathcal{D}(\mathcal{A})$. As in the classical case of $D[0,1]$, we shall see that it suffices to prove convergence of the finite dimensional distributions and tightness of the sequence.

First we deal with measurability. We introduce the following notation: for $A \in \mathcal{A}$,

$$[\emptyset, A] = \{A' \in \mathcal{A} : A' \subseteq A\}. \tag{7.9}$$

We observe that $[\emptyset, A] \times R \in \mathcal{G}$ whenever R is a compact set in $\mathcal{B}(\mathbf{R})$. We now need to consider the Borel sets on $(\mathcal{G}, d_G)$. We denote

$$\downarrow [\emptyset, A] = \{[\emptyset, A'] : A' \in \mathcal{A}, A' \subseteq A\}. \tag{7.10}$$

WEAK CONVERGENCE ON $\mathcal{D}(\mathcal{A})$ 145

We note that the class $\{\downarrow [\emptyset, A] : A \in \mathcal{A}\}$ is a π-system of sets (i.e., closed under finite intersection).

Similarly, for $R \in \mathcal{B}(\mathbf{R})$ ($\mathcal{B}(\mathbf{R})$ denotes the Borel sets of $\mathbf{R}$), $\downarrow R = \{R' \in \mathcal{B}(\mathbf{R}) : R' \subseteq R\}$.

Henceforth, whenever reference is made to the class $\mathcal{D}(\mathcal{A})$, it will be assumed that

Assumption 7.2.1 *The Borel sigma-algebra $\mathcal{B}_\mathcal{G}$ on $(\mathcal{G}, d_\mathcal{G})$ is generated by $\{\downarrow [\emptyset, A] \times \downarrow R : A \in \mathcal{A}, R \in \mathcal{B}(\mathbf{R})\}$.*

All of our examples satisfy this assumption.

The graph function $G : \mathcal{D}(\mathcal{A}) \to \mathcal{G}$ induces the σ-algebra $\mathcal{D} = G^{-1}(\mathcal{B}_\mathcal{G})$ on $\mathcal{D}(\mathcal{A})$. Since G is an isometry, it is easily seen that in fact $\mathcal{D}$ is the Borel σ-algebra on $(\mathcal{D}(\mathcal{A}), D_H)$.

Next, we examine the role of the finite dimensional projections. Given any $x \in \mathcal{D}(\mathcal{A})$, $n \in \mathbf{N}$ and any $A_1, \cdots, A_n \in \mathcal{A}$, define $\pi_{A_1, \cdots, A_n} : \mathcal{D}(\mathcal{A}) \to \mathbf{R}^n$ by

$$\pi_{A_1, \cdots, A_n}(x) = (x(A_1), \cdots, x(A_n)). \qquad (7.11)$$

Now, let $\mathcal{A}' = \cup_n \mathcal{A}_n$. $\mathcal{A}'$ is a countable d_H-dense subclass of $\mathcal{A}$. Let $\mathcal{D}' = \sigma(\{\pi_A : A \in \mathcal{A}'\})$ (the smallest σ-algebra on $\mathcal{D}(\mathcal{A})$ with respect to which each of the one-dimensional projections π_A is measurable). The following proposition shows that a $\mathcal{D}(\mathcal{A})$-valued process X is measurable if and only if X_A is a random variable for each $A \in \mathcal{A}'$.

Proposition 7.2.2 $\mathcal{D} = \mathcal{D}'$.

Proof. The proof is similar to that of Proposition 4.1 of Bass and Pyke (1985). It is easily seen that $\mathcal{D}' \subseteq \mathcal{D}$. To show the converse, by Assumption 7.2.1 it suffices to consider $G^{-1}(\downarrow [\emptyset, A] \times \downarrow R)$ for $A \in \mathcal{A}$ and $R \in \mathcal{B}(\mathbf{R})$. Now,

$$G^{-1}(\downarrow [\emptyset, A] \times \downarrow R) = \cap_{A' \in \mathcal{A}', A' \subseteq A} \{x \in \mathcal{D}(\mathcal{A}) : x(A') \in R\}.$$

This completes the proof. $\square$

It is now an immediate consequence of the preceding proposition that the finite-dimensional distributions are a determining class for probability measures on $\mathcal{D}(\mathcal{A})$:

Proposition 7.2.3 *Let μ_1 and μ_2 be any two probability measures supported on $(\mathcal{D}(\mathcal{A}), \mathcal{D})$. Then $\mu_1 = \mu_2$ if and only if*

$$\mu_1 \circ \pi_{A_1, \ldots, A_k}^{-1} = \mu_2 \circ \pi_{A_1, \ldots, A_k}^{-1}$$

for every $k \in \mathbf{N}$ and $A_1, \ldots, A_k \in \mathcal{A}'$.

We now turn to convergence of processes on $\mathcal{D}(\mathcal{A})$. Let $X, X_1, X_2,$ ... be $\mathcal{A}$-indexed processes taking their values in $\mathcal{D}(\mathcal{A})$, and let the corresponding probability measures induced on $\mathcal{D}(\mathcal{A})$ be denoted by $\mu, \mu_1, \mu_2, ...$, respectively. (Note: The processes do not need to be defined on the same probability space.) For notational convenience in what follows, we shall frequently write $X(A)$ instead of X_A.

Definition 7.2.4 (a) *The sequence (X_n) converges in finite dimensional distribution to X, denoted $X_n \to_{fdd} X$, if $(X_n(A_1), \cdots, X_n(A_m)) \to (X(A_1), \cdots, X(A_m))$ in distribution (as random vectors) for any $m \in \mathbf{N}$ and $A_1, \cdots, A_m \in \mathcal{A}$.*

(b) *$(X_n)_n$ is said to converge weakly in distribution on $\mathcal{D}(\mathcal{A})$ or functionally to X, denoted $X_n \to_\mathcal{D} X$, provided $\int f d\mu_n \to \int f d\mu$ for every function f on $\mathcal{G}$ which is bounded and continuous.*

Recall that a sequence of probability measures (μ_n) on $\mathcal{D}(\mathcal{A})$ is tight if for all $\epsilon > 0$ there exists a compact subset K_ϵ of $\mathcal{D}(\mathcal{A})$ such that $\inf_n \mu_n(K_\epsilon) \geq 1 - \epsilon$. Standard arguments give the following:

Theorem 7.2.5 *Let $X, X_1, X_2, \cdots$ be processes in $\mathcal{D}(\mathcal{A})$ with respective induced probability measures $\mu, \mu_1, \mu_2, ...$ on $\mathcal{D}(\mathcal{A})$. If*

(i) $X_n \to_{fdd} X$ *and*

(ii) (μ_n) *is tight,*

then $X_n \to_\mathcal{D} X$.

We may combine Theorems 7.1.15 and 7.2.5 to give the following generic functional limit theorem (cf. Slonowsky (1998)):

Theorem 7.2.6 *Suppose that $X, X_1, X_2, \cdots$ are processes in $\mathcal{D}(\mathcal{A})$ such that $X_n \to_{fdd} X$, and that for every $\epsilon > 0$ there exists a triple $(\Gamma_\epsilon, \Sigma_\epsilon, \Delta_\epsilon)$ as in Definition 7.1.10, satisfying the conditions of Theorem 7.1.15, such that*

$$P[X_n \in \Xi(\Gamma_\epsilon, \Sigma_\epsilon, \Delta_\epsilon)] \geq 1 - \epsilon,$$

then $X_n \to_\mathcal{D} X$.

We continue with a classic sufficient condition for functional convergence of sequences of processes with sample paths in $C(\mathcal{A})$.

Theorem 7.2.7 *(Slonowsky (1998))*
Let $X, X_1, X_2, \cdots$ be $\mathcal{A}$-indexed processes in $C(\mathcal{A})$. If

(i) $X_n \to_{fdd} X$,

SEMI-FUNCTIONAL CONVERGENCE

(ii) $\forall \eta > 0\ \exists a > 0$ such that $P[\,\|X_n\|_{\mathcal{A}} > a] < \eta\ \forall n$, and

(iii) $\forall \eta > 0$ and $\epsilon > 0\ \exists \delta > 0$ such that $P[w(X_n, \delta) \geq \epsilon] < \eta\ \forall n$

where $w(X_n, \delta) = \sup\{|X_n(A) - X_n(B)| : d_H(A,B) \leq \delta,\ A, B \in \mathcal{A}\}$, then $X_n \to_{\mathcal{D}} X$.

Proof. Since $C(\mathcal{A}) \subseteq \mathcal{D}(\mathcal{A})$, each X_n is in $\mathcal{D}(\mathcal{A})$. Let μ_n denote the probability measure induced on $\mathcal{D}(\mathcal{A})$ by X_n, $\forall n$. By Theorem 7.2.5 and assumption (i), it is sufficient to show (μ_n) is a tight family of measures on $\mathcal{D}(\mathcal{A})$.

Select $\eta > 0$. By (ii), $\exists a > 0$ such that

$$\mu_n(\{x \in C(\mathcal{A}) : \|x\|_{\mathcal{A}} \leq a\}) \geq 1 - \eta/2,\quad (\forall n) \qquad (7.12)$$

whereas (iii) implies that for every $k \in \mathbf{N}$, $\exists \delta_k > 0$ such that

$$\mu_n(\{x \in C(\mathcal{A}) : w(x, \delta_k) < 1/k\}) \geq 1 - \eta/2^{k+1},\quad (\forall n). \qquad (7.13)$$

If we define $\Sigma \subseteq C(\mathcal{A})$ by

$$\Sigma = \{x : \|x\|_{\mathcal{A}} \leq a\} \cap \cap_k \{x : w(x, \delta_k) < 1/k\},$$

then Σ is uniformly bounded and uniformly equicontinuous. Therefore, since $\mathcal{A}$ is d_H-compact, by the Arzelà-Ascoli characterization of compactness in $C(\mathcal{A})$ the $\|\cdot\|_{\mathcal{A}}$-closure, $\overline{\Sigma}$ of Σ is D_H-compact in $\mathcal{D}(\mathcal{A})$. Furthermore, by (7.12) and (7.13),

$$\mu_n(\overline{\Sigma}) \geq 1 - \eta,\quad (\forall n).$$

Since $\eta > 0$ was arbitrarily chosen, this implies (μ_n) is a tight family of probability measures on $\mathcal{D}$. □

We end this section with a brief comment about weak convergence of sequences of point processes. It is well-known that for sequences of simple point processes on $\mathbf{R}_+$, convergence in finite dimensional distribution to a limit which is also a simple point process is equivalent to weak convergence in the J_1 topology on $D[0, \infty)$. In the next chapter, an analogous result will be proven for set-indexed point processes in $\mathcal{D}(\mathcal{A})$. Thus, for sequences of point processes, it is often enough to consider only finite dimensional distributions, but it should be noted here that this is not true for general sequences in $\mathcal{P}(\mathcal{A})$.

7.3 Semi-Functional Convergence

One of the main goals of Part II of this book is to prove a central limit theorem (CLT) for sequences of set-indexed strong martingales. Ideally, this convergence would be functional convergence in $\mathcal{D}(\mathcal{A})$, as defined in the preceding section, to a set-indexed Brownian motion. Unfortunately, as we have already seen, if the class $\mathcal{A}$ is very large (for example, the lower sets in $[0,1]^2$), the Brownian motion may not be outer-continuous, and therefore cannot have a version with sample paths in $\mathcal{D}(\mathcal{A})$.

For this reason, we introduce a new mode of convergence, termed *semi-functional convergence*, which will be seen to be appropriate for set-indexed martingales. It includes a wide class of set-indexed processes, such as all strong martingales and many set-indexed Brownian motion processes.

We begin by returning to the concept of a *flow*, introduced in Chapter 5.

Definition 7.3.1 *A flow $f : [0,1] \to \mathcal{A}(u)$ is said to be simple provided that:*

1. *f is continuous,*
2. *$f(0) = \emptyset'$ and*
3. *for some $k \in \mathbf{N}$, there exist flows $f_i : [\frac{i-1}{k}, \frac{i}{k}] \to \mathcal{A}$, $i = 1, ..., k$ such that for $\frac{i-1}{k} \leq s \leq \frac{i}{k}$, $1 \leq i \leq k$,*

$$f(s) = f((i-1)/k) \cup f_i(s) = [\cup_{j=1}^{i-1} f_j(j/k)] \cup f_i(s).$$

The class of simple flows will be denoted by $S(\mathcal{A})$.

Remark 7.3.2 The class of simple flows is very rich: by a simple linear rescaling applied to the flow defined in Lemma 5.1.7, for any finite sub-semilattice $\mathcal{A}_1 = \{\emptyset' = A_0, ..., A_k\}$ of $\mathcal{A}$ equipped with a numbering consistent with the strong past, there exists a simple flow $f : [0,1] \to \mathcal{A}(u)$ such that $f(\frac{i}{k}) = \cup_{j=0}^{i} A_j$, $i = 0, ..., k$.

Definition 7.3.3 *The set $\mathcal{D}[S(\mathcal{A})]$ consists of all (additive) $\mathcal{A}$-indexed processes X such that $X \circ f$ has a modification in $D[0,1]$ for every $f \in S(\mathcal{A})$ (i.e., there exists a modification of $X \circ f$ which is right-continuous with left limits).*

$\mathcal{D}[S(\mathcal{A})]$ contains many important classes of $\mathcal{A}$-indexed processes, some of which are listed below.

Proposition 7.3.4 *Let X be an (additive) $\mathcal{A}$-indexed process. If*

(i) *X has purely atomic sample paths,*

(ii) *X is monotone inner- and monotone outer-continuous on $\mathcal{A}$,*

(iii) *X is a strong martingale, or*

(iv) X is a Brownian motion with monotone inner- and monotone outer- continuous variance measure Λ,

then $X \in \mathcal{D}[S(\mathcal{A})]$.

Proof. Fix a simple flow $f : [0,1] \to \mathcal{A}(u)$. Part (i) follows immediately, as $X \circ f$ is a pure jump process. Likewise, for (ii) we observe that $X \circ f$ is continuous. For (iii), by Lemma 5.1.2 it follows that $X \circ f$ is a (one-parameter) martingale, and so has a modification which is in $D[0,1]$. Finally, for (iv), since $\Lambda \circ f$ is continuous, there exists a continuous modification of the Gaussian process $X \circ f$. □

Now, for $X \in \mathcal{D}[S(\mathcal{A})]$ and $f \in S(\mathcal{A})$, we shall define $M_f(X)$ to be the unique (up to indistinguishability) modification of $X \circ f$ in $D[0,1]$.

Finally we introduce the concept of *semi-functional convergence*:

Definition 7.3.5 *Let* $X, X_1, X_2, \ldots \in \mathcal{D}[S(\mathcal{A})]$. *The sequence* (X_n) *converges semi-functionally to* X, *denoted* $X_n \to_{sf} X$, *if* $M_f(X_n)$ *converges weakly in* $D[0,1]$ *to* $M_f(X)$, *for every* $f \in S(\mathcal{A})$.

Comments 7.3.6 1. As in the previous section, we note that it is not necessary that the processes be defined on a common probability space.

2. Semi-functional convergence to a limit X such that $M_f(X)$ is continuous for every simple flow f implies convergence in finite dimensional distribution. This is a result of the following Proposition.

Proposition 7.3.7 *Let* $X, X_1, X_2, \ldots \in \mathcal{D}[S(\mathcal{A})]$. *If* $X_n \to_{sf} X$ *and* $M_f(X) \in C[0,1]$ *for every* $f \in S(\mathcal{A})$, *then* $(X_n(C_0), \ldots, X_n(C_m)) \to (X(C_0), \ldots, X(C_m))$ *in distribution (as random vectors) for any* $m \in \mathbf{N}$ *and* $C_1, \ldots, C_m \in \mathcal{C}(u)$. *Since* $\mathcal{A} \subseteq \mathcal{C}(u)$, *it follows that* $X_n \to_{sf} X$ *implies that* $X_n \to_{fdd} X$.

Proof. Without loss of generality, it may be assumed that the sets $C_0 = \emptyset', C_1, \ldots, C_m$ are the left-neighbourhoods of a finite sub-semilattice $\mathcal{A}' = \{A_0 = \emptyset', A_1, \ldots, A_m\}$ of $\mathcal{A}$ numbered in a manner consistent with the strong past. (A straightforward argument allows us to extend the proof to sets $C_i \in \mathcal{C}(u)$ (cf. Slonowsky (1998))). Then there exists $f \in S(\mathcal{A})$ such that $f(0) = C_0$ and for $1 \le i \le m$, $C_i = f(\frac{i}{m}) \setminus f(\frac{i-1}{m})$.

Since $M_f(X_n) \to M_f(X)$ weakly in $D[0,1]$ and $M_f(X)$ has continuous sample paths and since $M_f(X_n)(s) = X_n \circ f(s) = X_{n,f(s)}(:= X_n(f(s)))$ a.s., it follows that

$$(X_{n,f(0)}, X_{n,f(1/m)}, \ldots, X_{n,f(1)}) \to (X_{f(0)}, X_{f(1/m)}, \ldots, X_{f(1)})$$

in distribution. Now, since $X_n(C_i) = X_{n,f(i/m)} - X_{n,f((i-1)/m)}$ a.s., we may apply the continuous mapping theorem to obtain the result. □

Comment 7.3.8 Considering the case $T = [0,1]$ and $\mathcal{A} = \{[0,t] : t \in T\}$, we may identify $\mathcal{D}(\mathcal{A})$ with the usual function space $D[0,1]$. We recall that our definition of functional convergence is equivalent to weak convergence in the Skorokhod J_2 topology, while it is clear that semi-functional convergence is equivalent to weak convergence in the Skorokhod J_1 topology. As the J_1 topology is strictly stronger than the J_2 topology, it is clear that as defined here, functional convergence does not necessarily imply semi-functional convergence. Likewise, convergence of finite dimensional distributions does not imply weak J_1 convergence, so we do not have a converse to the preceding proposition. However, we do have a partial converse in the case that each X_n is a strong martingale and $M_f(X) \in C[0,1]$ for every f in $S(\mathcal{A})$.

Proposition 7.3.9 Let $X, X_1, X_2, \ldots \in \mathcal{D}[S(\mathcal{A})]$. Assume that

(i) (X_n) is a sequence of strong martingales,

(ii) for each $s \in [0,1]$ and $f \in S(\mathcal{A})$, $(X_n \circ f(s))_n$ is uniformly integrable,

(iii) $M_f(X) \in C[0,1]$ for every $f \in S(\mathcal{A})$.

Then $X_n \to_{sf} X \iff X_n \to_{fdd} X$.

Proof. In light of Proposition 7.3.7, it is only necessary to show that $X_n \to_{fdd} X$ implies that $X_n \to_{sf} X$. Let $f \in S(\mathcal{A})$. It is immediate that $X_n \to_{fdd} X$ implies that the finite dimensional distributions of $M_f(X_n)$ converge to those of $M_f(X)$. Thus, by Lemma 5.1.2 we have that $(M_f(X_n))_n$ is a sequence of martingales converging in finite dimensional distribution to the continuous limit $M_f(X)$ and (ii) implies that $(M_f(X_n)(s))_n$ is uniformly integrable for each $s \in [0,1]$. It was proven by Aldous (1989) that it follows that $(M_f(X_n))_n$ converges to $M_f(X)$ in $D[0,1]$. Therefore, $X_n \to_{sf} X$. □

CHAPTER 8

Limit Theorems for Point Processes

In this chapter, we apply results of the previous chapter to set-indexed point processes. It turns out that the space $\mathcal{D}(\mathcal{A})$ equipped with the topology as defined previously is particularly well-adapted to the study of weak convergence of point processes. Moreover, the study of point processes may be simplified by considering them as set-indexed submartingales and observing the behaviour of their compensators (see for example Ivanoff (1985), Brown, Ivanoff and Weber (1986) or Jacod and Shiryaev (1987)). Here too, the choice of the class $\mathcal{A}$ is very important. On the one hand, for the study of point processes the class must be sufficiently rich in order to generate a countable dissecting system, and a larger class $\mathcal{A}$ may admit more 'strictly simple' point processes. However, on the other hand, the smaller the class $\mathcal{A}$, the more likely it is that a sequence of point processes will have a corresponding sequence of compensators which are, in some sense, asymptotically deterministic.

In the first section, 'strictly simple' point measures are defined and the importance of their role is demonstrated. It turns out that the study of point processes on the line or the plane can be a bit misleading, in that a stronger definition of strict simplicity is needed in our general setting. The main result of the section is that, given a sequence of point measures and a strictly simple limit point measure, convergence in the vague topology is equivalent to convergence in $\mathcal{D}(\mathcal{A})$. The proof is not routine, but requires a careful study of the structure of strictly simple point processes. It is interesting to note that it is not enough to assume that the limiting point measure is 'simple' (i.e., there is no more than one jump at each point), even in the two parameter case (see for example Ivanoff (1985)). Next, we study weak convergence for set-indexed simple point processes. The main corollary is that if a sequence of set-indexed simple point processes converges in finite dimensional distribution to a set-indexed strictly simple point process, then the convergence is in the $\mathcal{D}(\mathcal{A})$ topology.

In the second section, we apply the results of the previous one to the Poisson process. We prove that, given a sequence of simple set-indexed point processes, if the associated sequence of compensators converges uniformly to a deterministic diffuse measure, then the simple point processes converge in finite dimensional distribution to a Poisson process. This result is a generalization of Ivanoff (1985), in which only the case $T = \mathbf{R}_+^2$ and $\mathcal{A}$ is the class of rectangles $[(0,0), z]$ is treated. However, even in this simple case, our result here is more general in that we do not require the limiting process to be strictly simple: we allow the Poisson process to be concentrated on any subset of T. The proof utilizes several ingredients: the constuction of a flow, a one-dimensional tightness condition due to Aldous (1978), the Watanabe characterization of a Poisson process on the line, and detailed calculations of various conditional probabilities. As a corollary, we obtain that if the limiting Poisson process is strictly simple, then the convergence is in $\mathcal{D}(\mathcal{A})$.

The last section is devoted to empirical processes and constitutes a nice statistical application of the weak convergence theorem and makes use of the different tools constructed above.

Most of the material presented here can be found in Ivanoff, Lin and Merzbach (1996) and for the last section see Ivanoff and Merzbach (1997).

8.1 Strictly simple point processes

We begin by defining a sub-class of the set-indexed purely atomic functions. As observed in the preceding chapter, we may assume without loss of generality that $T \in \mathcal{A}$ (or $\mathcal{A}(u)$). We also assume that $\mathcal{A}$ is consistent with boundaries and that Assumption 1.1.7 holds.

Definition 8.1.1 *Let $x \in \mathcal{D}(\mathcal{A})$. The function x is called a point measure if there exists a finite subset $T(x)$ of T such that $\forall A \in \mathcal{A}$, $x(A)$ is the cardinality of $A(x) = T(x) \cap A$. $A(x)$ is called the set of jump points of x in A.*

Remarks:
1. As the Borel sets $\mathcal{B}$ are generated by $\mathcal{A}$, x can be extended to a σ-additive measure on $\mathcal{B}$. Since $x(\{t\}) = 0$ or 1, $\forall t \in T$, x is often

STRICTLY SIMPLE POINT PROCESSES   153

called a simple point measure.
2. The topological space of (simple) point measures endowed with the metric D_H of $\mathcal{D}(\mathcal{A})$ is denoted $\mathcal{D}^1(\mathcal{A})$.

In order to motivate the results of this section, we recall some basic definitions and properties of point measures. First, we can define the *vague topology* on $\mathcal{D}^1(\mathcal{A})$:

Definition 8.1.2 *Let* $x, x_1, x_2, \ldots \in \mathcal{D}^1(\mathcal{A})$. *Then* x_n *converges to* x *vaguely in* $\mathcal{A}$ *if* $x_n(A) \to x(A)$ *as* $n \to \infty$ *for each* $A \in \mathcal{A}$ *such that* $x(\partial A) = 0$. *This will be denoted by* $x_n \to_v x$.

Comment 8.1.3 For any point measure x, a Borel set B is said to be a *continuity set* of x if $x(\partial B) = 0$. It is straightforward to show that $x_n \to_v x$ implies that $x_n(B) \to x(B)$ for each bounded set $B \in \mathcal{B}$ such that B is a continuity set of x. This is the usual definition of vague convergence of a sequence of measures (cf. Daley and Vere-Jones (1988), §A2.3). Indeed, it is enough to check this for the semi-algebra $\mathcal{C}$, which can be done by a simple approximation argument using continuity sets from $\mathcal{A}$.

The following examples clearly illustrate the differences between some of the topologies which can be defined on $\mathcal{D}^1(\mathcal{A})$:
1. Let $T = [0,2]^2$, and let $\mathcal{A}$ be the class of closed rectangles with the origin as a corner. Let $x_n = \delta_{(1/2, 1-1/n)} + \delta_{(1, 1+1/n)}$. This sequence converges to $x = \delta_{(1/2,1)} + \delta_{(1,1)}$ both in the vague topology and in the metric D_H, but not in the Skorokhod J_1 topology on $D([0,1]^2)$ (cf. Bickel and Wichura (1971)).
2. This example is similar, but now a third point is added to each measure: $x_n = \delta_{(1/2, 1-1/n)} + \delta_{(1, 1+1/n)} + \delta_{(1-1/n, 1/2)}$. This sequence converges in the vague topology to $x = \delta_{(1/2,1)} + \delta_{(1,1)} + \delta_{(1,1/2)}$, but now it does not converge in $\mathcal{D}(\mathcal{A})$. To see this, observe that for $A = [0,1]^2$, $x_n(A) = 2$ for all n, but $x(B) = 0, 1$ or $3, \forall B \subseteq T$. Also, $\{x_n\}$ does not converge in the Skorokhod J_1 topology.

For simple point measures on $\mathbf{R}_+$, it is well known that $x_n \to_v x$ is equivalent to convergence on $D[0,1]$, but the examples above show that this is no longer true in higher dimensions, nor is it equivalent to convergence on $\mathcal{D}(\mathcal{A})$. Therefore, our goal is to define a class of limiting point measures for which vague convergence is equivalent to convergence in the stronger topology generated by D_H.

Before stating the main result of this section, some natural assumptions are needed on the class $\mathcal{A}$.

We will assume first that an equivalent metric (also denoted by d) exists on T such that for any $A \in \mathcal{A}$ and $\epsilon > 0$, the set $A^\epsilon = \{t \in T : d(t, A) \leq \epsilon\} \in \mathcal{A}$. We make this assumption for clarity of exposition only: an assumption analogous to (A1)(iii) of Bass and Pyke (1985) would suffice. We shall also assume that:

Assumption 8.1.4 Let $A_t = \cap_{A \in \mathcal{A}, t \in A} A$. Then for every $\epsilon > 0$, there exists $n_\epsilon < \infty$ such that if $n \geq n_\epsilon$ then for every $t \in T$, $diam(g_n(A_t) \setminus \cup_{A \in \mathcal{A}_n, t \notin A} A) < \epsilon$ $((diam(C) = \sup_{s,t \in C} d(s, t))$.

Assumption 8.1.7 following forces the class $\mathcal{A}$ to be large enough to be able to separate ('shatter') points of certain finite subsets of T. First, we make this concept precise.

Definition 8.1.5 For any subclass $\mathcal{A}'$ of $\mathcal{A}$, and any finite subset S of T, we say that S is shattered by $\mathcal{A}'$ if the number of different sets $A \cap S$ for $A \in \mathcal{A}'$ is equal to 2^n, where n is the cardinality of S.

Comment:
This definition means simply that for any subset S' of S, one can find a set $A \in \mathcal{A}'$ such that $A \cap S = S'$.

Definition 8.1.6 Two finite subsets R and S of T are called ϵ-twins if for each point $s \in S$ there exists a unique point $r \in R$ such that $d(s, r) < \epsilon$, and vice versa.

Assumption 8.1.7 Let R and S be two finite subsets of T and suppose that $\mathcal{H} \subseteq \mathcal{A}$ is a class such that $\mathcal{H}$ shatters S and $R \subseteq B, \forall B \in \mathcal{H}$. Then there exists $\epsilon^*(R, S) > 0$ such that for all $\epsilon \leq \epsilon^*(R, S)$, if R' and S' are ϵ-twins of R and S, respectively, then there exists a class $\mathcal{H}'$ such that $\mathcal{H}'$ shatters S', $R' \subseteq B', \forall B' \in \mathcal{H}'$ and $\forall B' \in \mathcal{H}'$, $\exists B \in \mathcal{H}$ such that $d_H(B, B') < \epsilon$.

Henceforth, we suppose that Assumption 8.1.7 holds. It is easily seen to be true for our examples.

We are now ready to define a strictly simple point measure.

Definition 8.1.8 A point measure x defined on T is called strictly simple if there exists a positive number $\rho(x)$ such that for any $0 \leq \rho < \rho(x)$, any $\epsilon > 0$ and any $A \in \mathcal{A}$, the set $(\partial A)^\rho(x)$ is shattered by the class $\mathcal{A}' = \{A' \in \mathcal{A} : d_H(A, A') < \rho + \epsilon\}$ (∂A denotes the boundary of A).

Remarks:
1. For $\rho = 0$, this means that for any subset S of $(\partial A)(x)$ and any $\epsilon > 0$, there exists a set $A' \in \mathcal{A}$ such that $d_H(A, A') < \epsilon$ and

STRICTLY SIMPLE POINT PROCESSES

$(\partial A)(x) \cap A' = S$.

2. When $T = [0,1]^2$ and $\mathcal{A}$ is the usual class of rectangles, Definition 8.1.8 implies that for any strictly simple point measure x, for every jump point t of x, $x(\partial A_t) = 1$. This is the intuitive notion of a strictly simple point measure, and this property plus the requirement that there be no jumps on the axes are in fact equivalent to Definition 8.1.8.

3. In more general spaces, Definition 8.1.8 is strictly stronger than the intuitive notion. For example, if $T = [0,1]^d, d \geq 3$, and $\mathcal{A}$ is the usual class of rectangles, a point measure is strictly simple if and only if there is at most one jump point on each hyperplane parallel to each of the $(d-1)$-dimensional subspaces determined by $(d-1)$ of the axes and there are no jumps on the axes. In the more complicated example of the function space introduced in the first chapter, two functions f and g cannot both be jump points of a strictly simple point measure if either $f \leq g$ and $f(t) = g(t)$ for some $t \in T$, or f and g attain their sups at the same point.

We have the following lemma:

Lemma 8.1.9 *If x is a strictly simple point measure, then $\forall \rho < \rho(x)$, $\forall A \in \mathcal{A}$, $\forall \epsilon > 0$ $\exists$ a class $\mathcal{H} \subseteq \mathcal{A}$, such that :*
(i) $(A\backslash(\partial A)^\rho)(x) \subseteq B$, $\forall B \in \mathcal{H}$.
(ii) $\mathcal{H}$ shatters $(\partial A)^\rho(x)$.
(iii) $d_H(A,B) < \rho + \epsilon$, $\forall B \in \mathcal{H}$.

Proof. Since $(\partial A)^\rho$ is closed and $(A\backslash(\partial A)^\rho)(x)$ is finite, there exists $\epsilon > 0$ such that $d(t,(\partial A)^\rho) > \epsilon, \forall t \in (A\backslash(\partial A)^\rho)(x)$. Let $\mathcal{H} = \{B \in \mathcal{A} : d_H(A,B) < \rho + \epsilon\}$. By Definition 8.1.8, (ii) is satisfied, and by the choice of ϵ, (i) is satisfied. □

The main result of this section is the following theorem from Ivanoff, Lin and Merzbach (1996):

Theorem 8.1.10 *Given a sequence (x_n) of point measures and a strictly simple point measure x, convergence of $(x_n(A))$ to $x(A)$ on all continuity sets $A \in \mathcal{A}$ of x is equivalent to convergence of (x_n) to x in $\mathcal{D}(\mathcal{A})$. In other words,*

$$x_n \to_v x \iff x_n \to_{D_H} x.$$

Proof. We begin by noting that convergence of $\{x_n\}$ to x on continuity sets is equivalent to requiring that for any $A \in \mathcal{A}$ such that $x(\partial A) = 0$, there exists N (which may depend on A) such that $x_n(A) = x(A)$, $\forall n \geq N$.

Convergence of $\{x_n\}$ to x in $\mathcal{D}(\mathcal{A})$ implies that for every $\epsilon > 0$, there exists N such that $\forall A \in \mathcal{A}$ and $n \geq N$, there exist sets $A', A'' \in \mathcal{A}$ (which may depend on n) with $d_H(A, A') < \epsilon$, $d_H(A, A'') < \epsilon$, and $x(A) = x_n(A')$, and $x_n(A) = x(A'')$.

First, we assume that $\{x_n\}$ converges to x in $\mathcal{D}(\mathcal{A})$. Let A be a continuity set of x. Then $0 < \delta = \min_{t \in T(x)}\{d(t, \partial A)\}$. Thus, if $B \in \mathcal{A}$ is such that $d_H(A, B) < \delta$, then $x(A) = x(B)$. If $\{x_n(A)\}$ does not converge to $x(A)$, then there exists a subsequence $\{x_{n'}(A)\}$ such that either $x_{n'}(A) < x(A) \;\forall n'$ or $x_{n'}(A) > x(A) \;\forall n'$. Suppose that $x_{n'}(A) < x(A) \;\forall n'$. Then for every $B \in \mathcal{A}$ such that $d_H(A, B) < \delta$, $x(A) = x(B) > x_{n'}(A)$, contradicting the assumption that $\{x_n\}$ converges to x in $\mathcal{D}(\mathcal{A})$. The proof is analogous in the case that $x_{n'}(A) > x(A) \;\forall n'$. Thus, by contradiction, $\{x_n(A)\}$ converges to $x(A)$ for all continuity sets $A \in \mathcal{A}$ of x.

Conversely, assume now that $\{x_n(A)\}$ converges to $x(A)$ for all continuity sets $A \in \mathcal{A}$ of x. The proof consists of four claims, whose proofs will follow.

Claim 1: Given x and $\{x_n\}$, for every $\epsilon > 0$ sufficiently small, $\exists n_\epsilon$ such that whenever $n \geq n_\epsilon$, $T(x)$ and $T(x_n)$ are ϵ-twins.

Claim 2: Given x and $\{x_n\}$, if $\rho(x)$ is as in Definition 8.1.8, then for any $A \in \mathcal{A}$, $\rho < \rho(x)$ and $\epsilon < \rho/2$ sufficiently small, if $n \geq n_\epsilon$ then $(\partial A)^\epsilon(x_n)$ is shattered by a class $\mathcal{H}_n$ such that $(A \backslash (\partial A)^\epsilon)(x_n) \subseteq B$, $B \subseteq A^\epsilon$ and $d_H(A, B) < 4\epsilon$, $\forall B \in \mathcal{H}_n$.

Claim 3: Let y and z be two point measures such that $y(T)$ and $z(T)$ are ϵ-twins. Suppose that, for any $A \in \mathcal{A}$, there exists a class $\mathcal{H}_y^A$ ($\mathcal{H}_z^A$) shattering $(\partial A)^\epsilon(y)$ $((\partial A)^\epsilon(z))$ such that $(A \backslash (\partial A)^\epsilon)(y) \subseteq B$, $((A \backslash (\partial A)^\epsilon)(z) \subseteq B)$, $B \subseteq A^\epsilon$, and $d_H(A, B) < 4\epsilon$, $\forall B \in \mathcal{H}_y^A$ ($\mathcal{H}_z^A$). Then $\forall A \in \mathcal{A}$, $\exists A', A'' \in \mathcal{A}$ such that $y(A) = z(A')$, $z(A) = y(A'')$, and $d_H(A, A') < 4\epsilon$ and $d_H(A, A'') < 4\epsilon$.

Claim 4: For every $\epsilon > 0$, there exists n_ϵ such that $D_H(x_n, x) < 5\epsilon$, $\forall n \geq n_\epsilon$.

Proof of Claim 1:

Let $t_1, ..., t_k$ be the jump points of x and let $d = \min_{i \neq j} d(t_i, t_j)$. For $\epsilon < d/4$, by Assumption 8.1.4 (iii), there is $m_\epsilon < \infty$ such that for $m \geq m_\epsilon$ and for each i, $1 \leq i \leq k$, t_i is an interior point of $C_{m,i} = g_m(A_{t_i}) \backslash \cup_{A \in \mathcal{A}_m, t_i \notin A} A$ and diam $(C_{m,i}) \leq \epsilon$, so $C_{m,i}$ is a continuity set of x containing exactly one jump point of x. Thus, if $m \geq m_\epsilon$ is fixed, there exists n_ϵ such that if $n \geq n_\epsilon$, then $x(T) = x_n(T) = k$ and $x(C_{m,i}) = x_n(C_{m,i})$, $\forall i = 1, ..., k$. In particular, this implies, for $n \geq n_\epsilon$, that x_n has k jump points, that for each jump point t_i of x, there exists a jump point t'_i of x_n such that $d(t_i, t'_i) < \epsilon$, and that the jump point t'_i is unique. Clearly, the converse is true as well, proving Claim 1.

Proof of Claim 2:

Since the number of points in x is finite, $\exists \epsilon^* > 0$, such that for all sets of points of the form $S = (\partial A)^\delta(x)$ and $R = (A \backslash (\partial A)^\delta)(x)$, for $\rho(x) > \delta > 0$ and $A \in \mathcal{A}$, $\epsilon^*(R,S) \geq \epsilon^*$ ($\epsilon^*(R,S)$ is defined in Assumption 8.1.7). Now fix $\epsilon > 0$ such that $\epsilon < \rho/2, \epsilon < \epsilon^*/2$ and ϵ is sufficiently small that Claim 1 is satisfied. Let $n \geq n_\epsilon$ and let $A \in \mathcal{A}$. Consider $S_n = (\partial A)^\epsilon(x_n)$ and $R_n = (A \backslash (\partial A)^\epsilon)(x_n)$, and let S_x and R_x be the ϵ-twins of S_n and R_n defined by x. Then $S_x \subseteq (\partial A)^{2\epsilon}(x)$, and as well, $R_x \subseteq A$. If we let $S'_x = (S_x \cup R_x) \cap (\partial A)^{2\epsilon}(x)$ and $R'_x = R_x \backslash S'_x$, then by Lemma 8.1.9, there exists a class $\mathcal{H}$ shattering S'_x such that $R'_x \subseteq B$, and $d_H(A, B) < 3\epsilon \; \forall B \in \mathcal{H}$.

Now let S'_n and R'_n be the ϵ-twins of S'_x and R'_x defined by x_n. Clearly, $S_n \cup R_n = S'_n \cup R'_n$, $S_n \subseteq S'_n$, and $R'_n \subseteq R_n$. By Assumption 8.1.7, there exists a class $\mathcal{H}'_n$ shattering S'_n such that $R'_n \subseteq B$, and $d_H(A, B) < 4\epsilon \; \forall B \in \mathcal{H}'_n$. Since $S_n \subseteq S'_n$, it is clear that there is a subclass $\mathcal{H}_n$ of $\mathcal{H}'_n$, each set of which contains R_n and which shatters S_n.

Proof of Claim 3:

Fix $A \in \mathcal{A}$. By symmetry, it suffices to prove the existence of A'. Consider $A(y)$ and let Q_z be the ϵ-twins of $A(y)$ defined by z (note: $Q_z \neq A(z)$ in general). If $s \in A(y)$, then its twin s' in z is in $A^\epsilon(z)$. On the other hand, if $s' \in (A^\epsilon \backslash (\partial A)^\epsilon)(z) = (A \backslash (\partial A)^\epsilon)(z)$, then its twin s in y is in $A(y)$. Therefore, $Q_z \subseteq A^\epsilon(z)$, and if $s' \in A^\epsilon(z) \backslash Q_z$, then $s' \in (\partial A)^\epsilon(z)$. By hypothesis, there exists $A' \in \mathcal{A}$ such that $A' \subseteq A^\epsilon$, $d_H(A, A') < 4\epsilon$, $Q_z \subseteq A'$, and $(A^\epsilon(z) \backslash Q_z) \cap A' = \emptyset$. This implies that $A'(z) = Q_z$, as required.

Proof of Claim 4:

To prove Claim 4, note that $D_H(x_n, x) = d_G(G(x_n), G(x))$, where $G(x)$ is the closure of $\{(A, x(A)) : A \in \mathcal{A}\}$. Thus, since x is a point measure, the points in $G(x)$ are of the form $(A, x(A))$ or $(A, \lim_i x(B_i))$, where $A, B_i \in \mathcal{A}$, $d_H(A, B_i) \to 0$, and $(x(B_i))$ is ultimately constant. The same comments are true for each point measure x_n as well.

We shall assume, without loss of generality, that ϵ is small enough that Claims 1 and 2 are valid. Using the notation of Claim 2, therefore, for $n \geq n_\epsilon$, x_n and x both satisfy the conditions of Claim 3.

Then, by Claim 3, for any $n \geq n_\epsilon$ and any $A \in \mathcal{A}$, sets $A', A'' \in \mathcal{A}$ may be found such that $d_H(A, A') < 4\epsilon$, $d_H(A, A'') < 4\epsilon$ and $x(A) = x_n(A')$, $x_n(A) = x(A'')$. Thus, whenever $n \geq n_\epsilon$, for points in $G(x)$ $(G(x_n))$ of the form $(A, x(A))$ $((A, x_n(A)))$, there exists a point $(A', x_n(A'))$ in $G(x_n)$ $((A'', x(A''))$ in $G(x))$ such that $\rho((A, x(A)), (A', x_n(A'))) < 4\epsilon$ $(\rho((A, x_n(A)), (A'', x(A''))) < 4\epsilon)$. For points in $G(x)$ of the form $(A, \lim_i x(B_i))$, choose j sufficiently large that $d_H(A, B_j) < \epsilon$ and that $x(B_j) = \lim_i x(B_i)$. Then for $n \geq n_\epsilon$, there exists a point $(B, x_n(B))$ in $G(x_n)$ such that $\rho((B_j, x(B_j)), (B, x_n(B))) < 4\epsilon$. Finally, we obtain

$$\rho((A, \lim_i x(B_i)), (B, x_n(B)))$$
$$\leq \rho((A, \lim_i x(B_i)), (B_j, x(B_j))) + \rho((B_j, x(B_j)), (B, x_n(B)))$$
$$< 5\epsilon.$$

A similar argument shows that for a point in $G(x_n)$ of the form $(A, \lim_i x_n(B_i))$, if $n \geq n_\epsilon$, there exists a point $(B, x(B))$ in $G(x)$ such that

$$\rho((A, \lim_i x_n(B_i)), (B, x(B))) < 5\epsilon.$$

Thus, $D_H(x_n, x) < 5\epsilon$, for $n \geq n_\epsilon$, completing the proof of Claim 4 and of Theorem 8.1.10. □

Definition 8.1.11 *Let X be an integrable $\mathcal{A}$-indexed process which is adapted to the filtration $\{\mathcal{F}_A : A \in \mathcal{A}\}$. Assume that $X_{\emptyset'} = 0$ and that each sample path $X(\omega)$ of X has a finitely additive extension to $\mathcal{B}$.*
(i) Any Borel set A is called a stochastic continuity set of X *if $P[X(\partial A) > 0] = 0$.*
(ii) X is called a simple point process *if each realization $X(\omega)$, $\omega \in \Omega$, belongs to $\mathcal{D}^1(\mathcal{A})$.*
(iii) X is called a strictly simple point process *if each realization $X(\omega)$, $\omega \in \Omega$, is a strictly simple point measure.*

We shall need the following lemma.

Lemma 8.1.12 *Let M be an integrable (not necessarily adapted) process of bounded variation (i.e. a difference of two increasing processes) such that $E[M_C \mid \mathcal{G}_C^*] = 0 \ \forall C \in \mathcal{C}$. Then if $B, B' \in \tilde{\mathcal{A}}(u)$, $B \subseteq B'$,*

$$E[M_{B'} - M_B \mid \mathcal{F}_B] = 0.$$

Proof. Since M is of bounded variation, M_B and $M_{B'}$ are well-defined. First assume that $B, B' \in \mathcal{A}(u)$. Let $B' = \cup_{i=1}^k A_i$, $A_i \in$

STRICTLY SIMPLE POINT PROCESSES   159

A, $i = 1, \ldots, k$. Without loss of generality, assume that for each $i, j \in \{1, \ldots, k\}$, there exists $\ell \in \{1, \ldots, k\}$ such that $A_i \cap A_j = A_\ell$. (This representation is not extremal.) Then we may write $B' \setminus B = \cup_{i=1}^k A_i \setminus B = \cup_{i=1}^k \left(A_i \setminus \cup_{j: A_i \not\subseteq A_j} A_j \cup B \right) = \cup_{i=1}^k C_i$, where $C_1, \ldots, C_k$ are disjoint sets in $\mathcal{C}$. Clearly, $\mathcal{F}_B \subseteq \mathcal{G}_{C_i}^*$, $i = 1, \ldots, k$. Thus,

$$E(M_{B'} - M_B | \mathcal{F}_B) = E\left(\sum_{i=1}^k M_{C_i} \Big| \mathcal{F}_B \right)$$

$$= E\left(\sum_{i=1}^k E(M_{C_i} | \mathcal{G}_{C_i}^*) \Big| \mathcal{F}_B \right) = 0.$$

Now consider the general case: $B, B' \in \tilde{\mathcal{A}}(u)$.

$E(M_{B'} - M_B | \mathcal{F}_B)$
$= \lim_n E\left(M_{g_n(B')} - M_{g_n(B)} \big| \mathcal{F}_B \right)$
(by outer − continuity and uniform integrability)
$= \lim_n \lim_{m \geq n} E\left(M_{g_n(B')} - M_{g_n(B)} \big| \mathcal{F}_{g_m(B)} \right)$
(by the reverse martingale convergence theorem)
$= 0$ (since $g_m(B) \subseteq g_n(B)$ if $m \geq n$). $\square$

Recall that '$\to_{\mathcal{D}}$' denotes weak convergence in the topology generated on $\mathcal{D}^1(\mathcal{A})$ by the metric D_H. We shall use the notation '$\xrightarrow{\mathcal{D}}$' for convergence in distribution of other sequences of random elements, and the relevant topology will be indicated.

The following theorem from Ivanoff, Lin and Merzbach (1996) will be seen to be a direct consequence of Theorem 8.1.10. We note that for point processes, $X_n \to X$ in distribution in the vague topology on $\mathcal{D}^1(\mathcal{A})$ when X is strictly simple if and only if the finite dimensional distributions of (X_n) converge to those of X on stochastic continuity sets of X (cf. Daley and Vere-Jones (1988)). By the finite additivity of X and the continuous mapping theorem, the class of stochastic continuity sets for the vague convergence can be chosen to be either from $\mathcal{A}$ or $\mathcal{C}$.

Theorem 8.1.13 *Suppose that (X_n) is a sequence of simple point processes and that X is a strictly simple point process. If Assump-*

tion 8.1.4 holds, then $X_n \xrightarrow{\mathcal{D}} X$ in the vague topology if and only if $X_n \longrightarrow_{\mathcal{D}} X$.

Proof. Let $\mathcal{S}$ denote the event 'X not strictly simple'. We shall show that $\mathcal{S}$ is measurable and that $P_n(\mathcal{S}) \to 0$, where P_n is the measure induced by X_n. Since any function continuous in $\mathcal{D}^1(\mathcal{A})$ is also continuous in the vague topology when restricted to $\mathcal{S}^c$, the theorem follows.

Measurability of $\mathcal{S}$ is straightforward: for m arbitrary but fixed, any realization of X which is not strictly simple is contained in a finite union of events of the form $\cap_{k \in \mathcal{I}} \{X(C_k^m) > 0\}$, where $\mathcal{I}$ is some finite index set, and the sets C_k^m are disjoint 'left-neighbourhoods' in $\mathcal{C}_m$ (i.e., of the form $C_k^m = A \setminus \cup_{A' \in \mathcal{A}_m; A \not\subseteq A'} A'$ for some $A \in \mathcal{A}_m$). Denote this finite union of events by $\mathcal{S}_m$. Since $\{\mathcal{C}_m\}_m$ separates points of T, we have that $\mathcal{S} = \cap_m \mathcal{S}_m$. Thus, for any $\epsilon > 0$, m can be chosen so that $P(\mathcal{S}_m) < \epsilon$ (P is the measure induced by X). If it can be assumed that all the sets $A \in \cup_m \mathcal{A}_m$ are continuity sets, then by convergence of finite dimensional distributions, $\lim_n P_n(\mathcal{S}_m) < \epsilon$, and so $\lim_n P_n(\mathcal{S}) = 0$, since ϵ is arbitrary. But it is easily shown that the sets $A \in \mathcal{A}$ which are stochastic continuity sets are dense in the Hausdorff topology in $\mathcal{A}$. Thus, the semilattices $\mathcal{A}_m$ may be replaced in the above argument by semilattices $\mathcal{A}'_m$ which consist of stochastic continuity sets, which satisfy the property of 'separability from above', and which generate 'left-neighbourhoods' which separate points in T. This completes the proof. □

8.2 Poisson limit theorem

We shall apply the preceding results to the Poisson process. Note that we still assume that T is compact and that $\mathcal{A}$ is consistent with boundaries.

Before presenting the Poisson limit theorem, note that we shall endow the function space $D[0, L]$ ($L < \infty$) with Skorokhod J_1-topology. For typographical reasons, in what follows subscripts may appear instead in parentheses: $X_C = X(C)$.

We shall make use of Aldous' well-known stopping time condition which gives an elegant sufficient condition for tightness on the function space $D[0, L]$:

Theorem 8.2.1 *(Aldous (1978))*
Let (Y_n) be a sequence of random elements of $D[0, L]$. Then (Y_n)

is tight in the Skorokhod J_1-topology on $D[0,L]$ if:
(i) $(Y_n(s))_n$ is tight on the line for each $s \in [0,L]$, and
(ii) $Y_n(\tau_n + \delta_n) - Y_n(\tau_n) \to_P 0$ for each sequence (τ_n, δ_n) such that τ_n is a natural stopping time for Y_n $\forall n$, $\delta_n \in \mathbf{R}_+$, $\forall n$ and $\delta_n \to 0$.

Theorem 8.2.2 and its corollary are from Ivanoff, Lin and Merzbach (1996):

Theorem 8.2.2 *Let (X_n) be a sequence of simple point processes, and let (V_n) be a corresponding sequence of *-compensators. Suppose that $(X_n(T))$ and $(V_n(T))$ are uniformly integrable and that there exists a deterministic finite diffuse measure V such that*

$$\sup_{A \in \tilde{\mathcal{A}}(u)} | V(A) - V_n(A) | \to_P 0.$$

Then for any finite collection of sets $C_1, ... C_k \in \mathcal{C}$, $k \in \mathbf{N}$, the sequence of joint distributions of $(X_n(C_1), ..., X_n(C_k))$ converges to the joint distribution of $(X(C_1), ..., X(C_k))$ as $n \to \infty$, where X is a Poisson process with mean measure V.

Proof. We begin by outlining the main idea of the proof:

To show convergence of finite dimensional distributions for sets in $\mathcal{C}$, it will be shown that if $C_1, ... C_k \in \mathcal{C}$ are disjoint, then $X_n(C_1), ..., X_n(C_k)$ converge in distribution to independent Poisson random variables with means $V(C_1), ..., V(C_k)$, respectively.

We do this by associating to each X_n a (not necessarily simple) point process N_n defined on a bounded subinterval $[0,L]$ of $\mathbf{R}_+$ such that $X_n(C_i)$ may be expressed as

$$X_n(C_i) = \sum_{j_i=1}^{h_i} N_n(I_{i,j_i}),$$

where (I_{i,j_i}) are disjoint (left-open, right-closed) intervals, $j_i = 1, ..., h_i$, $i = 1, ..., k$. (Note: the same intervals are used for each n.) We then show that $N_n \xrightarrow{\mathcal{D}} N$ in the Skorokhod J_1-topology on $[0,L]$, where N is a Poisson process with mean measure U, and

$$V(C_i) = \sum_{j_i=1}^{h_i} U(I_{i,j_i}).$$

Thus, by convergence of the finite dimensional distributions of (N_n)

and the continuous mapping theorem,

$$(X_n(C_1), ..., X_n(C_k)) = (\sum_{j_1=1}^{h_1} N_n(I_{1,j_1}), ..., \sum_{j_k=1}^{h_k} N_n(I_{k,j_k}))$$

$$\xrightarrow{D} (\sum_{j_1=1}^{h_1} N(I_{1,j_1}), ..., \sum_{j_k=1}^{h_k} N(I_{k,j_k}))$$

$$= (X(C_1), ..., X(C_k)),$$

where $X(C_1), ..., X(C_k)$ are independent Poisson random variables with means $V(C_1), ..., V(C_k)$, respectively. (In the above expression, '$\xrightarrow{D}$' denotes the usual convergence in distribution of a random vector in $\mathbf{R}^k$.)

We now define the (one-dimensional) processes (N_n) and N. We may assume without loss of generality that all of the sets in $\mathcal{A}$ which define the sets $C_1, ...C_k$ are included in $\mathcal{A}_1$. We denote by $\mathcal{C}^\ell(\mathcal{A}_1)$ the class of left-neighbourhoods generated by the sets in $\mathcal{A}_1$. Let $\mathcal{A}_1 = \{\emptyset' = A_0, ..., A_j\}$ be a numbering consistent with the strong past. We may define as in Chapter 6 a flow $\theta : [0, j] \to \tilde{\mathcal{A}}(u)$ such that

(i) $\theta(i) = \cup_{h=0}^{i} A_h$.
(ii) Each $C \in \mathcal{C}^\ell(\mathcal{A}_1)$ is of the form $C = \theta(i) \setminus \theta(i-1), i = 1, ..., j$
(iii) $U(\cdot) = V(\theta(\cdot))$ is a continuous deterministic increasing function on $[0, j]$.

Let $N_n(N)$ be defined by $N_n(s) = X_n(\theta(s))(N(s) = X(\theta(s)))$. Similarly, $U_n(U)$ is defined by $U_n(s) = V_n(\theta(s))(U(s) = V(\theta(s)))$. Then clearly for each n, $X_n(C_i)$ can be expressed as a finite sum

$$X_n(C_i) = \sum_{h \in \mathcal{I}_i} X_n(\theta(h) \setminus \theta(h-1)) = \sum_{h \in \mathcal{I}_i} N_n(h-1, h],$$

where $\mathcal{I}_i$ is some finite index set. Likewise,

$$X(C_i) = \sum_{h \in \mathcal{I}_i} N(h-1, h].$$

Now we will show that in the Skorokhod J_1-topology on $D[0, j]$, $N_n \xrightarrow{D} N$, where N is a Poisson process with mean measure U. This gives the required finite dimensional convergence. The proof that $N_n \xrightarrow{D} N$ will be carried out in three steps:

Step 1: Let $\mathcal{H}_n(s) = \sigma(N_n(u) : u \leq s)$. It will be shown that for

POISSON LIMIT THEOREM 163

any $\mathcal{H}_n$-stopping times τ, σ with $\tau < \sigma$,

$$E(N_n(\tau) - U_n(\tau) - (N_n(\sigma) - U_n(\sigma)) \mid \mathcal{H}_n(\sigma)) = 0.$$

Note: U_n is not necessarily $\mathcal{H}_n$-adapted.

Step 2: We use the result in Step 1 and the one-dimensional stopping time condition of Theorem 8.2.1 to show that (N_n) is tight (in $D[0,j]$). Let N be any weak limit. The uniform integrability of (X_n) and (V_n) is used to show that $N - U$ is a martingale with respect to the minimal filtration generated by N. Clearly, N is an increasing process taking its values in $\mathbf{Z}_+$.

Step 3: It will be shown that N is a simple point process on the line.

Then by Watanabe's characterization theorem for the Poisson process on the line, N is necessarily a Poisson process with mean measure U, completing the proof of the theorem.

We proceed with the proofs of the three steps:

Proof of step 1:

We note first that $\mathcal{H}_n$ is right-continuous (see for example, Liptser and Shiryayev (1978), Lemma 18.4) and that $\mathcal{H}_n(s) \subseteq \mathcal{F}_n(\theta(s))$ $\forall s$. Let $M_n = N_n - U_n$, and assume initially that σ and τ take on finitely many values, say $0 = s_0 < s_1 < ... < s_r$. Let $F \in \mathcal{H}_n(\sigma)$. Then,

$$\int_F M_n(\tau) - M_n(\sigma)\, dP = E(\sum_{i=1}^{r} I_{(\sigma < s_i \leq \tau)}(M_n(s_i) - M(s_{i-1}))I_F).$$

Noting that $I_{(\sigma < s_i \leq \tau)} I_F$ is $\mathcal{H}_n(s_{i-1})$-measurable since $F \in \mathcal{H}_n(\sigma)$,

$$\int_F M_n(\tau) - M_n(\sigma)\, dP$$
$$= E(\sum_{i=1}^{r} E((M_n(s_i) - M_n(s_{i-1})) \mid \mathcal{H}_n(s_{i-1}))I_{(\sigma < s_i \leq \tau)}I_F)$$
$$= E(\sum_{i=1}^{r} E(E((X_n - V_n)(\theta(s_i) \backslash \theta(s_{i-1}))$$
$$\quad \mid \mathcal{F}_n(\theta(s_{i-1}))) \mid \mathcal{H}_n(s_{i-1}))I_{(\sigma < s_i \leq \tau)}I_F)$$
$$= 0.$$

The last equality follows by Lemma 8.1.12.

Since $\mathcal{H}_n$ is right-continuous, this result may be extended to arbitrary stopping times in the usual way by approximating from above using discrete stopping times. This completes the proof of step 1.

Proof of step 2:

We first verify the conditions of Theorem 8.2.1. Let $\{\tau_n\}$ be a sequence of $\mathcal{H}_n$-stopping times, and let $\{\delta_n\}$ be any sequence of positive constants such that $\delta_n \to 0$. Therefore

$$\begin{aligned}
| U_n(\tau_n + \delta_n) - U_n(\tau_n) | & \\
= \quad & | V_n(\theta(\tau_n + \delta_n)) - V_n(\theta(\tau_n)) | \\
\leq \quad & 2 \sup_{A \in \tilde{A}(u)} | V(A) - V_n(A) | + U(\tau_n + \delta_n) - U(\tau_n) \\
\to_P \quad & 0.
\end{aligned}$$

By uniform integrability of $\{V_n(T)\}$,

$$\begin{aligned}
E(N_n(\tau_n + \delta_n) - N_n(\tau_n)) &= E(U_n(\tau_n + \delta_n) - U_n(\tau_n)) \\
&\to 0.
\end{aligned}$$

Thus, the conditions of Theorem 8.2.1 are satisfied. Uniform integrability now implies that $\{N_n\}$ is tight in $D[0, j]$, and if N is any weak limit, that $N - U$ is a martingale with respect to its minimal filtration.

Proof of step 3:

It remains to show that any weak limit N (in $D[0, j]$) of $\{N_n\}$ is a simple point process on the line. The technique used in this proof is a refinement of that used in the proof of Theorem 5.3.1.

We fix $n \geq 0$, and define $S_0^n, ..., S_{2^n}^n$ as in Theorem 5.3.1: $S_0^n = 0$, $S_{2^n}^n = j$, and if $1 \leq i \leq 2^n - 1$, $S_i^n = \inf\{s \geq S_{i-1}^n : U_s - U_{S_{i-1}^n} = U(j)/2^n\}$. Thus, $U_{S_i^n} = ia/2^n, i = 0, ..., 2^n$, where $a = U(j)$.

Let N be any weak limit of N_n in $D[0, j]$. Now, it is easily seen that

$$\begin{aligned}
\{\omega : N(\omega) \text{ not simple}\} & \\
= \quad & \{\omega : \exists s \in [0, j] \text{ such that } N(s) - N(s-) > 1\} \\
\subseteq \quad & \cap_n \mathcal{U}_n,
\end{aligned}$$

where $\mathcal{U}_n = \cup_{i=1}^{2^n} \{N(S_i^n \backslash S_{i-1}^n) > 1\}$. We shall show that $P(\mathcal{U}_n) \to 0$ as $n \to \infty$.

Since the finite dimensional distributions of N_m converge to those of N,

$$P(\mathcal{U}_n) = \lim_{m \to \infty} P(\cup_{i=1}^{2^n} \{N_m(S_i^n \backslash S_{i-1}^n) > 1\}). \tag{8.1}$$

POISSON LIMIT THEOREM   165

Now,

$$P(\cup_{i=1}^{2^n}\{N_m(S_i^n\setminus S_{i-1}^n) > 1\})$$

$$= \sum_{i=1}^{2^n} P[\{N_m(S_i^n\setminus S_{i-1}^n) > 1\} \cap (\cap_{j\leq i-1}\{N_m(S_j^n\setminus S_{j-1}^n) \leq 1\})]$$

$$= \sum_{i=1}^{2^n} P[\{N_m(S_i^n\setminus S_{i-1}^n) > 1\} \cap \mathcal{V}_m^n(i-1)], \qquad (8.2)$$

where $\mathcal{V}_m^n(i-1) = \cap_{j\leq i-1}\{N_m(S_j^n\setminus S_{j-1}^n) \leq 1\}$. Note that $\mathcal{V}_m^n(i-1) \in \mathcal{H}_m(S_{i-1}^n) \subseteq \mathcal{F}_m(\theta(S_{i-1}^n))$.

We partition the set $\theta(S_i^n)\setminus\theta(S_{i-1}^n)$ exactly as in the proof of Theorem 5.3.1. For $r > n$, let $\mathcal{A}_{n,r}$ be the class of all intersections of sets in $\mathcal{A}_r \cup \{\theta(S_i^n); 0 \leq i \leq 2^n\}$ and let $\mathcal{A}_{n,r}(u)$ be the class of (finite) unions of sets in $\mathcal{A}_{n,r}$. Let $\mathcal{L}_{n,r}$ be the class of 'left-neighbourhoods' of sets in $\mathcal{A}_{n,r}$:

$$\mathcal{L}_{n,r} = \{C : C = B\setminus \cup_{B'\in\mathcal{A}_{n,r}, B\not\subseteq B'} B', B \in \mathcal{A}_{n,r}\}.$$

Define for $i = 1,...,2^n$,

$$\mathcal{L}_{n,r}(i) = \{C \in \mathcal{L}_{n,r} : C \subseteq \theta(S_i^n)\setminus\theta(S_{i-1}^n)\}.$$

We have that $\theta(S_i^n)\setminus\theta(S_{i-1}^n) = \cup_{C\in\mathcal{L}_{n,r}(i)} C$ and $\mathcal{L}_{n,r} = \cup_{i=1}^{2^n}\mathcal{L}_{n,r}(i)$, and these unions are disjoint. Any set $C \in \mathcal{L}_{n,r}(i)$ is of the form $A\setminus B$, $A \in \mathcal{A}_{n,r}$, $B \in \mathcal{A}_{n,r}(u)$. The sets of $\mathcal{L}_{n,r}(i)$ (say $C_1,...,C_q$) may be ordered so that if $C_j = A_j\setminus B_j$ and $C_h = A_h\setminus B_h$ and $A_h \subset A_j$, then $h < j$. Let '$\prec$' denote the induced total order on $\mathcal{L}_{n,r}(i)$ (i.e. $C \prec C' \Leftrightarrow C = C_h, C' = C_j$ where $h < j$). For $C \in \mathcal{L}_{n,r}(i)$, let $E_C = \cup_{C':C\prec C'} C'$. It is easily seen that E_C has the form $E_C = \theta(S_i^n)\setminus B_C$, for some $B_C \in \tilde{\mathcal{A}}(u)$ such that $\theta(S_{i-1}^n) \subseteq B_C$ and $C' \subseteq B_C$ if $C' \prec C$. Define:

$$\mathcal{D}_C^m = \{X_m(C) = 1\} \cap \cap_{C'\prec C}\{X_m(C') = 0\}$$

$$\mathcal{E}_C^m = \{X_m(E_C) \geq 1\}$$

$$\mathcal{Q}_{n,r}^m(i) = \{\omega : \exists C \in \mathcal{L}_{n,r}(i) \text{ such that } X_m(C) > 1\}.$$

We observe that $\mathcal{D}_C^m \in \mathcal{F}_m(B_C)$, and that the sets $\mathcal{D}_C^m, \mathcal{D}_{C'}^m$ are disjoint if $C \neq C'$.

Now, if $X_m(\theta(S_i^n)\setminus\theta(S_{i-1}^n))(\omega) > 1$, then either one of the sets in $\mathcal{L}_{n,r}(i)$ contains more than one point, in which case $\omega \in \mathcal{Q}_{n,r}^m(i)$, or there exist two sets $C, C' \in \mathcal{L}_{n,r}(i)$ with $C \prec C'$ and $X_m(C) =$

$X_m(C') = 1$, in which case $\omega \in \cup_{C \in \mathcal{L}_{n,r}(i)}(\mathcal{D}_C^m \cap \mathcal{E}_C^m)$. Thus,

$$\{N_m(S_i^n \backslash S_{i-1}^n) > 1\}$$
$$= \{X_m(\theta(S_i^n) \backslash \theta(S_{i-1}^n)) > 1\}$$
$$\subseteq \{\cup_{C \in \mathcal{L}_{n,r}(i)}(\mathcal{D}_C^m \cap \mathcal{E}_C^m)\} \cup \mathcal{Q}_{n,r}^m(i)\}.$$

Therefore,

$$P(\{N_m(S_i^n \backslash S_{i-1}^n) > 1\} \cap \mathcal{V}_m^n(i-1))$$
$$\leq \sum_{C \in \mathcal{L}_{n,r}(i)} P(\mathcal{D}_C^m \cap \mathcal{E}_C^m \cap \mathcal{V}_m^n(i-1)) + P(\mathcal{Q}_{n,r}^m(i))$$
$$= \sum_{C \in \mathcal{L}_{n,r}(i)} P(\mathcal{E}_C^m \mid \mathcal{D}_C^m \cap \mathcal{V}_m^n(i-1)) P(\mathcal{D}_C^m \cap \mathcal{V}_m^n(i-1))$$
$$+ P(\mathcal{Q}_{n,r}^m(i)).$$

Since $\mathcal{D}_C^m \cap \mathcal{V}_m^n(i-1) \in \mathcal{F}_m(B_C)$,

$$P(\mathcal{E}_C^m \mid \mathcal{D}_C^m \cap \mathcal{V}_m^n(i-1)) \leq E(N_m(E_C) \mid \mathcal{D}_C^m \cap \mathcal{V}_m^n(i-1))$$
$$= E(V_m(E_C) \mid \mathcal{D}_C^m \cap \mathcal{V}_m^n(i-1))$$
$$= \frac{E(V_m(E_C) I(\mathcal{D}_C^m \cap \mathcal{V}_m^n(i-1)))}{P(\mathcal{D}_C^m \cap \mathcal{V}_m^n(i-1))}.$$

Thus,

$$P(\{N_m(S_i^n \backslash S_{i-1}^n) > 1\} \cap \mathcal{V}_m^n(i-1))$$
$$\leq \sum_{C \in \mathcal{L}_{n,r}(i)} E(V_m(E_C) I(\mathcal{D}_C^m \cap \mathcal{V}_m^n(i-1))) + P(\mathcal{Q}_{n,r}^m(i))$$
$$\leq \sum_{C \in \mathcal{L}_{n,r}(i)} E(V(E_C) I(\mathcal{D}_C^m \cap \mathcal{V}_m^n(i-1)))$$
$$+ \sum_{C \in \mathcal{L}_{n,r}(i)} E(|V_m(E_C) - V(E_C)| I(\mathcal{D}_C^m \cap \mathcal{V}_m^n(i-1)))$$
$$+ P(\mathcal{Q}_{n,r}^m(i))$$
$$\leq \frac{a}{2^n} P(\mathcal{V}_m^n(i-1))$$
$$+ \sum_{C \in \mathcal{L}_{n,r}(i)} E(|V_m(E_C) - V(E_C)| I(\mathcal{D}_C^m \cap \mathcal{V}_m^n(i-1)))$$
$$+ P(\mathcal{Q}_{n,r}^m(i)). \tag{8.3}$$

The last inequality above follows by the fact that $E_C \subseteq \theta(S_i^n) \backslash \theta(S_{i-1}^n)$, and the events $\mathcal{D}_C^m$ are disjoint.

Consider the second term in the expression above. We have that

$|V_m(E_C) - V(E_C)| \leq 2\sup_{A \in \tilde{\mathcal{A}}(u)} |V_m(A) - V(A)| = Y_m$ (say), and $Y_m \to_P 0$. Also, $\{Y_m\}$ is uniformly integrable by the uniform integrability of $\{V_m(T)\}$. Thus, given $\epsilon > 0$, we may choose $\alpha < \infty$ such that $E(Y_m I(Y_m > \alpha)) < \epsilon, \forall m$. Next choose m large enough that $P(Y_m > \epsilon) < \epsilon/\alpha$. Thus,

$$\sum_{C \in \mathcal{L}_{n,r}(i)} E(|V_m(E_C) - V(E_C)| I(\mathcal{D}_C^m \cap \mathcal{V}_m^n(i-1)))$$

$$\leq \sum_{C \in \mathcal{L}_{n,r}(i)} [E(Y_m I(Y_m \leq \epsilon) I(\mathcal{D}_C^m \cap \mathcal{V}_m^n(i-1)))$$
$$+ E(Y_m I(\epsilon < Y_m \leq \alpha) I(\mathcal{D}_C^m \cap \mathcal{V}_m^n(i-1)))$$
$$+ E(Y_m I(\alpha < Y_m) I(\mathcal{D}_C^m \cap \mathcal{V}_m^n(i-1)))]$$

$$\leq \sum_{C \in \mathcal{L}_{n,r}(i)} [\epsilon P(\mathcal{D}_C^m \cap \mathcal{V}_m^n(i-1))$$
$$+ \alpha P((Y_m > \epsilon) \cap (\mathcal{D}_C^m \cap \mathcal{V}_m^n(i-1)))$$
$$+ E(Y_m I(\alpha < Y_m) I(\mathcal{D}_C^m \cap \mathcal{V}_m^n(i-1)))]$$

$$\leq \epsilon P(\mathcal{V}_m^n(i-1)) + \alpha P((Y_m > \epsilon) \cap \mathcal{V}_m^n(i-1))$$
$$+ E(Y_m I(Y_m > \alpha) I(\mathcal{V}_m^n(i-1))), \qquad (8.4)$$

since the events $\mathcal{D}_C^m$ are disjoint. Finally, combining equations (8.2)-(8.4) and noting that the events $\mathcal{V}_m^n(i), \mathcal{V}_m^n(j)$ are disjoint if $i \neq j$,

$$P(\cup_{i=1}^{2^n} N_m(S_i^n \setminus S_{i-1}^n) > 1)$$
$$= \sum_{i=1}^{2^n} P(\{N_m(S_i^n \setminus S_{i-1}^n) > 1\} \cap \mathcal{V}_m^n(i-1))$$
$$\leq \left(\frac{a}{2^n} + \epsilon\right) \sum_{i=1}^{2^n} P(\mathcal{V}_m^n(i-1)) + \alpha \sum_{i=1}^{2^n} P((Y_m > \epsilon) \cap \mathcal{V}_m^n(i-1))$$
$$+ \sum_{i=1}^{2^n} E(Y_m I(Y_m > \alpha) I(\mathcal{V}_m^n(i-1))) + \sum_{i=1}^{2^n} P(\mathcal{Q}_{n,r}^m(i))$$
$$\leq \frac{a}{2^n} + \epsilon + \alpha P(Y_m > \epsilon) + E(Y_m I(Y_m > \alpha)) + \sum_{i=1}^{2^n} P(\mathcal{Q}_{n,r}^m(i)).$$

However, r is arbitrary, so we may let $r \to \infty$ in the above expression (holding m and n fixed). Since $\{\mathcal{C}_r\}_r$ separates points, as

$r \to \infty$

$$P(\mathcal{Q}_{n,r}^m(i)) \to P\{X_m \text{ is not simple on } \theta(S_i^n)\backslash\theta(S_{i-1}^n)\} = 0.$$

Therefore,

$$\begin{aligned}
P(\cup_{i=1}^{2^n} N_m(S_i^n \backslash S_{i-1}^n) > 1) \\
\leq \frac{a}{2^n} + \epsilon + \alpha P(Y_m > \epsilon) + E(Y_m I(Y_m > \alpha)) \\
\leq \frac{a}{2^n} + 3\epsilon \quad (8.5)
\end{aligned}$$

for all m sufficiently large.

Thus, since ϵ is arbitrary, we may substitute (8.5) in (8.1) and let $m \to \infty$ to obtain

$$P(\mathcal{U}_n) \leq \frac{a}{2^n}.$$

Finally, let $n \to \infty$ to show that any weak limit N in $D[0, j]$ of $\{N_n\}$ is simple with probability one: $P\{\omega : N(\omega) \text{ not simple}\} = \lim_n P(\mathcal{U}_n) = 0$.

This completes the proof of the theorem. $\square$

Theorems 8.1.13 and 8.2.2 allow us to prove convergence to a Poisson process in $D^1(\mathcal{A})$, provided that the limit is strictly simple.

Corollary 8.2.3 *If X is strictly simple and if the conditions of Theorem 8.2.2 are satisfied, then $X_n \to_D X$.*

Theorem 8.2.2 can be extended to doubly stochastic Poisson processes (Cox processes) (see Definition 3.4.7 and Proposition 3.4.8) as in Cojocaru and Merzbach (1999):

Theorem 8.2.4 *Let (X_n) be a sequence of simple point processes and let (V_n) be a corresponding sequence of *-compensators. Suppose that $(X_n(T))$ and $(V_n(T))$ are uniformly integrable and that there exists an increasing and continuous process V such that*
(i) $\sup_{A \in \tilde{\mathcal{A}}(u)} |V(A) - V_n(A)| \xrightarrow{P} 0$, and
(ii) for every $n = 1, 2, \ldots$, V is $\mathcal{G}_{\emptyset'}^{*n}$-measurable (the weak past of $\emptyset'$, associated with the process X_n).
Then (X_n) converges in distribution to a doubly stochastic Poisson process directed by V.

The proof is essentially the same as that of Theorem 8.2.2 and therefore we omit it.

8.3 Empirical processes

Consider an experiment whose outcome is an $\mathbf{R}^d$-valued random variable, repeated countably many times, giving rise to a family $\{Y_i\}_{i=1}^{\infty}$ of independent identically distributed random variables. The set-indexed empirical process associated with observing the first n of the $\{Y_i\}$ is the process

$$X_n(A) = \frac{1}{n} \sum_{i=1}^{n} I_{\{Y_i \in A\}},$$

where A belongs to an indexing collection $\mathcal{A}$ in $\mathbf{R}^d$ and $I(\cdot)$ denotes the indicator function. The main result of this section is that if the distribution function of Y is continuous and 'differentiable' at a point, then the rescaled process $nX_n(A/n^{1/d})$ converges in distribution to a set-indexed Poisson process whose rate depends on the derivative of the distribution function. These results illustrate the ease with which the theory of set-indexed martingales may be applied to certain statistical problems.

In order to exploit martingale theory, a set-indexed compensator is required, and we have computed such a compensator for a single jump process in Section 4.5. We use it to find the compensator of the rescaled empirical process; the compensator is then shown to converge uniformly to a deterministic measure absolutely continuous with respect to Lebesgue measure. By Theorem 8.2.2, we conclude that the rescaled empirical process converges to a Poisson process.

When the random variables are $\mathbf{R}_+$-valued, these results were obtained by Al-Hussaini and Elliott (1984), and we are able to use a similar method of proof. However, even in this special case our result is more general in that we do not require that the random variables be non-negative as the set-indexed framework gives us a kind of two-sided martingale.

These results come from Ivanoff and Merzbach (1997).

Assume that $T = \prod_{i=1}^{d}[-K, K]$, for some $K \in \mathbf{N}$. We recall the following notation, introduced in Section 4.5. For any point $t = (t_1, ..., t_d) \in T$, let $R_t = \prod_{i=1}^{d} \triangle t_i$, where

$$\triangle t_i = \begin{cases} [0, t_i] & \text{if } t_i > 0 \\ \{0\} & \text{if } t_i = 0 \\ [t_i, 0] & \text{if } t_i < 0 \end{cases}$$

Also, $S_t = T \cap \prod_{i=1}^{d} \triangle' t_i$, where

$$\Delta' t_i = \begin{cases} \mathbf{R}_+ \setminus \Delta t_i & \text{if } t_i > 0 \\ \mathbf{R}_- \setminus \Delta t_i & \text{if } t_i < 0. \end{cases}$$

S_t is not defined if $t_i = 0$ some $i, 1 \leq i \leq d$.

We assume that our indexing collection $\mathcal{A}$ is either the class of rectangles R_t or the collection of all the lower sets of T. In any case, we can consider sets in one or more quadrants simultaneously.

Now, let Y be a T-valued random variable and let F be its distribution function: $F(t) = P(Y \leq t), t \in T$. The set-indexed single jump process X associated with Y is defined by $X_A = I_{\{Y \in A\}}, A \in \mathcal{A}$. If $\{\mathcal{F}_A\}$ be the minimal filtration generated by X, we recall that a *-compensator of X is $\tilde{X}$, where

$$\tilde{X}_A = \int_{A \cap R_Y} (F(\overline{S_u}))^{-1} dF(u). \tag{8.6}$$

(We use the convention that $(F(\overline{S_u}))^{-1} = 0$ if $F(\overline{S_u}) = 0$ or if $u_i = 0$ for some $i, 1 \leq i \leq d$.)

Let $Y_1, Y_2, \ldots$ be independent and identically distributed T-valued random variables, with continuous distribution function F. The empirical measure associated with $Y_1, \ldots, Y_n$ is

$$X_n(A) = \frac{1}{n} \sum_{i=1}^{n} I_{\{Y_i \in A\}},$$

where A is any Borel set. Restricting our attention to sets in $\mathcal{A}$, we define the set-indexed point process

$$N_n(A) = nX_n(A/n^{1/d}) = \sum_{i=1}^{n} I_{\{Y_i \in A/n^{1/d}\}},$$

where $A/n^{1/d} = \{t = (t_1, \ldots, t_d) \in \mathbf{R}^d : (n^{1/d}t_1, \ldots, n^{1/d}t_d) \in A\}$. We note that all processes here have sample paths which are in $\mathcal{D}(\mathcal{A})$.

The following is a reformulation of Corollary 8.2.3, noting that N is a Poisson process on $\mathbf{R}^d$ with mean measure V which is absolutely continuous with respect to λ, and therefore that N is strictly simple.

Theorem 8.3.1 *Let (N_n) be a sequence of $\mathcal{A}$-indexed simple point processes and let (V_n) be a corresponding sequence of *-compensators. Suppose that $(N_n(T))$ and $(V_n(T))$ are uniformly integrable and that there exists a deterministic measure V absolutely contin-*

EMPIRICAL PROCESSES   171

uous with respect to λ such that

$$\sup_{A \in \tilde{\mathcal{A}}(u)} |V(A) - V_n(A)| \to_P 0. \tag{8.7}$$

Then $N_n \to_D N$, where N is a Poisson process with mean measure V.

We shall now apply the previous theorem to the empirical measure. Let $Q = D_1 \times \ldots \times D_d$ where $D_i = (0, \infty)$ or $(-\infty, 0)$, $i = 1, \ldots, d$ denote an (open) quadrant in $\mathbf{R}^d$.

Definition 8.3.2 *When it exists, the derivative of the measure F at 0 from the quadrant Q is $F'_Q(0)$, where*

$$F'_Q(0) = \lim_{t \to_Q 0} \frac{F(R_t)}{|\prod_{i=1}^d t_i|}.$$

Here, '$t \to_Q 0$' means that $t_i \to 0$, $i = 1, \ldots, d$ and $t \in Q, \forall t$.

Theorem 8.3.3 *Let the $\mathcal{A}$-indexed process N_n be defined as above and suppose that $F'_Q(0)$ exists for each quadrant $Q \subset \mathbf{R}^d$. Then as $n \to \infty$, $N_n \to_D N$, where N is a Poisson process with mean measure V which is 0 on all of the $(d-1)$-dimensional hyperplanes $\{t = (t_1, \ldots, t_d) \in T : t_i = 0\}$, $i = 1, \ldots, d$ and which is equal to $F'_Q(0)\lambda(\cdot)$ on each open quadrant $Q \cap T$ of T.*

Proof. The conditions of Theorem 8.3.1 will be verified. Let $\mathcal{F}_A = \sigma\{I_{\{Y_i \in A'\}} : i \in \mathbf{N}, A' \subseteq A, A' \in \mathcal{A}\} \vee \mathcal{F}_0$, where $\mathcal{F}_0$ denotes the P-null sets. By independence of $Y_1, Y_2, \ldots$, from (8.6) we see that the *-compensator of N_n with respect to the filtration $\mathcal{F} = \{\mathcal{F}_A : A \in \mathcal{A}\}$ is

$$\tilde{N}_n(A) = \sum_{i=1}^n \int_{R_{Y_i} \cap A/n^{1/d}} (F(\overline{S_u}))^{-1} dF(u).$$

It is easily seen that $E(N_n(A)) = E(\tilde{N}_n(A)) = nF(A/n^{1/d})$, and that $E((N_n(A))^2) = nF(A/n^{1/d}) + n(n-1)(F(A/n^{1/d}))^2$. We note that for $A = R_t$, $F(A) = F'_Q(0)\lambda(A) + o(\lambda(A))$, and so $nF(A/n^{1/d}) \to F'_Q(0)\lambda(A)$. Thus the random variables $(N_n(T))$ are uniformly integrable. We shall show that for each $A = R_t$,

$$\tilde{N}_n(A) \to_{L^2} F'_Q(0)\lambda(A). \tag{8.8}$$

This in turn gives uniform integrability of $(\tilde{N}_n(T))$ and convergence in probability of the compensator on rectangles R_t. The fact that all sets in $\tilde{\mathcal{A}}(u)$ may be uniformly approximated by finite unions of such rectangles permits us to extend Lemma 1 of McLeish (1978) to prove (8.7), and so the theorem follows.

To complete the proof, we shall show that $E((\tilde{N}_n(A))^2) \to (F'_Q(0)\lambda(A))^2$ for $A = R_t$, thereby proving (8.8). Now,

$$(\tilde{N}_n(A))^2 = \sum_{i=1}^{n} \left[\int_{R_{Y_i} \cap A/n^{1/d}} (F(\overline{S_u}))^{-1} dF(u) \right]^2 \quad (8.9)$$

$$+ \sum_{i \neq j} \left[\int_{R_{Y_i} \cap A/n^{1/d}} (F(\overline{S_u}))^{-1} dF(u) \right]$$

$$\cdot \left[\int_{R_{Y_j} \cap A/n^{1/d}} (F(\overline{S_v}))^{-1} dF(v) \right]. \quad (8.10)$$

By independence, the expected value of (8.10) is

$$n(n-1)(F(A/n^{1/d}))^2 \to (F'_Q(0)\lambda(A))^2.$$

We shall show that the expected value of (8.9) converges to 0.

To avoid trivialities, assume that $F(Q) > 0$. Let $t \in Q$, $Y \sim F$ and let $f(t) = \int_{R_t} I_{\{u \in R_Y\}}(F(\overline{S_u}))^{-1} dF(u)$. We have the following integration by parts formula:

$$\int_{R_t} f(R_u) df(u) = \int_{R_t} f(S_u) df(u)$$

where $f(R_u)$ and $f(S_u)$ denote the increments of f over R_u and S_u, respectively (cf. Hildebrandt (1963)). Thus, $(f(t))^2$ may be expressed as a linear combination of integrals of the form $\int_{R_t} f(u') df(u)$, where u' is such that $u'_i = u_i \ \forall i \in J$, where J is some nonempty subset of $\{1, ..., d\}$, and $u_i = t_i \ \forall i \notin J$. Now,

$$E\left[\int_{R_t} f(u') df(u) \right] = \int_{R_t} \int_{R_{u'}} \frac{F(\overline{S_v} \cap \overline{S_u})}{F(\overline{S_u}) F(\overline{S_v})} dF(v) dF(u)$$

$$\leq \int_{R_t} \int_{R_{u'}} (F(\overline{S_u}))^{-1} dF(v) dF(u)$$

$$= \int_{R_t} (F(\overline{S_u}))^{-1} F(R_{u'}) dF(u)$$

$$\leq (F(\overline{S_t}))^{-1} (F(R_t))^2.$$

EMPIRICAL PROCESSES

Thus, there exists a finite constant c such that

$$\limsup_n E((f(t/n^{1/d}))^2)$$
$$\leq c\limsup_n (F(\overline{S_t/n^{1/d}}))^{-1}(F(R_t/n^{1/d}))^2$$
$$\leq c(F(Q))^{-1}\limsup_n [F'_Q(0)\lambda(A)/n + o(1/n)]^2.$$

Substituting in (8.9),

$$E\left(\sum_{i=1}^n \left[\int_{R_{Y_i}\cap A/n^{1/d}} (F(\overline{S_u}))^{-1} dF(u)\right]^2\right) = nE((f(t/n^{1/d}))^2) \to 0.$$

This completes the proof of the theorem. □

CHAPTER 9

Martingale Central Limit Theorems

In this important final chapter we will present semi-functional and functional central limit theorems (CLT's) for set-indexed strong martingales. There is an abundant literature about central limit theorems for random functions indexed by subsets of metric spaces, but most of these works require a metric entropy condition (see, for example, Dudley (1984) or Ossiander (1987)). Here we take a different approach, and use set-indexed martingale tools. The results presented in this chapter are the culmination of the work in the area to date, and open the door to the development of a martingale-based theory of estimation and inference for set-indexed processes. We shall see that the theorems are very general, with easily verified conditions, and in fact generalize the classical martingale CLT's on $\mathbf{R}_+$.

In Section 9.1, we present both a semi-functional and a functional central limit theorem. As we shall see, the conditions closely mimic the usual martingale CLT's on $\mathbf{R}_+$: the sequence of martingales must have asymptotically deterministic quadratic variation processes, and must also satisfy a condition of 'asymptotic rarefaction of jumps'. As an example, in Section 9.2 we will show that the set-indexed weighted empirical process satisfies the conditions of a martingale CLT. The results here are based primarily on Slonowsky (1998) and Slonowsky and Ivanoff (1999).

9.1 Central Limit Theorems

We begin with a semi-functional central limit theorem. As noted in Chapter 7, *we may assume without loss of generality that T is compact and that $T \in \mathcal{A}$ (or $\mathcal{A}(u)$)*, since both semi-functional and functional convergence follow from the asymptotic behaviour of the sequence of processes on compact subsets of T.

Before presenting the central limit theorem, we shall recall some notation and important concepts from the theory of the function space $D[0,1]$.

First, we shall always endow $D[0,1]$ with the Skorokhod J_1-topology. Next, if $y \in D[0,1]$, let $J(y) = \sup_{0 \leq s \leq 1} |y(s) - y(s-)|$. J is known as the *jump functional* on $D[0,1]$. The following result is Proposition 3.26 of Jacod and Shiryaev (1987):

Proposition 9.1.1 *Let (Y_n) be a sequence of random elements of $D[0,1]$. Assume that (Y_n) is tight in the Skorokhod J_1-topology on $D[0,1]$ and that $J(Y_n) \to_P 0$ as $n \to \infty$. If some subsequence $Y_{n'} \to Y$ weakly in $D[0,1]$, then $Y \in C[0,1]$ (i.e., almost all sample paths of Y are continuous).*

We now consider a simple flow $f : [0,1] \to \mathcal{A}(u)$ (cf. Definition 7.3.1). Recall that any strong martingale X is in the class $\mathcal{D}[S(\mathcal{A})]$ - i.e., there exists a modification of $X \circ f$ in $D[0,1]$ - and that $M_f(X)$ denotes the right-continuous modification of $X \circ f$. Recall that an increasing process Q is a *-quadratic variation for X if for every $C \in \mathcal{C}$,
$$E[X_C^2 \mid \mathcal{G}_C^*] = E[Q_C \mid \mathcal{G}_C^*].$$

We are ready to state the semi-functional central limit theorem presented in Slonowsky (1998) and Slonowsky and Ivanoff (1999), but the proof given here is different. To avoid multiple subscripts in what follows, we shall frequently replace $(X_n)_A$ with $X_n(A)$, etc.

Theorem 9.1.2 *Let $(X_n, \mathcal{F}_n, P_n)$ be a sequence of strong martingales defined on $\mathcal{A}$, and let (Q_n) be any sequence of corresponding *-quadratic variation processes. Assume that*
1. $\sup_n E[|X_n(T)|^{2+\delta}] < \infty$ *for some* $\delta > 0$.
2. $J(M_f(X_n)) \to_P 0$ *as* $n \to \infty$ *for every simple flow* f.
3. $\{Q_n(T)\}$ *is uniformly integrable.*
4. *For every* $A \in \mathcal{A}$, $Q_n(A) \to_P \Lambda(A)$, *where* Λ *is a deterministic, increasing, monotone inner- and outer-continuous function on* $\mathcal{A}$.

Then there exists a Brownian motion X defined on $\mathcal{A}$ with variance function Λ such that
$$X_n \to_{sf} X.$$

Comment 9.1.3 As stated in the introduction to this chapter, the preceding theorem is new even for martingales on $\mathbf{R}_+$, in that there is no requirement that the quadratic variation processes be predictable or even adapted.

CENTRAL LIMIT THEOREMS   177

Proof. It must be shown that $M_f(X_n)$ converges weakly in $D[0,1]$ to $M_f(X)$, for every $f \in S(\mathcal{A})$ (the class of simple flows). First we shall show that for $f \in S(\mathcal{A})$ fixed but arbitrary,

(I) $(M_f(X_n))_n$ *is tight in* $D[0,1]$.

By assumption (2) and Proposition 9.1.1, any weak limit Y of $(M_f(X_n))_n$ must be continuous. Next, we shall prove that

(II) *Any weak limit* Y *of* $(M_f(X_n))_n$ *is a martingale with quadratic variation* $\Lambda \circ f$.

Therefore, since Y is continuous, by Lévy's characterization Y is a Brownian motion process on $[0,1]$ with variance measure $\lambda = \Lambda \circ f$. Finally, since Y has the same finite-dimensional distributions as $M_f(X)$, $M_f(X_n)$ converges weakly to $M_f(X)$ in $D[0,1]$. Since $f \in S(\mathcal{A})$ was arbitrary,

$$X_n \to_{sf} X.$$

It remains to prove (I) and (II) above. We may assume without loss of generality that $f(1) = T$, since by Lemma 5.1.7 any simple flow may be extended to include T. Denote $Y_n = M_f(X_n)$ and $\lambda_n = Q_n \circ f$. Since f is a simple flow taking its values in $\mathcal{A}(u)$, by Lemma 5.1.2, Y_n is a martingale with respect to the filtration $\mathcal{H}_n = \{\mathcal{H}_n(s) = \mathcal{F}_n(f(s)) : 0 \leq s \leq 1\}$.

Proof of (I):
The conditions of Theorem 8.2.1 will be verified for the sequence $(M_f(X_n)) = (Y_n)$. First, we observe that $(|Y_n|^{2+\delta})$ is a submartingale, and so $E[|Y_n(s)|^{2+\delta}] \leq E[|Y_n(1)|^{2+\delta}] = E[|X_n(T)|^{2+\delta}]$ $\forall s$, $0 \leq s \leq 1$. Therefore, $(M_f(X_n)(s)))_n = (Y_n(s))_n$ is tight in $\mathbf{R}$ by the uniform bound on $E[|X_n(T)|^{2+\delta}]$. Next, as observed by Aldous (1978), by right continuity of Y_n, it is enough to verify condition (ii) of Theorem 8.2.1 for stopping times τ_n taking on countably many values. Therefore, for any sequence (τ_n) of natural stopping times taking on at most countably many values $(s_{n,1}, s_{n,2}, ...)$, and any sequence (δ_n) such that $\delta_n \to 0$ as $n \to \infty$,

$$E[(Y_n(\tau_n + \delta_n) - Y_n(\tau_n))^2] \qquad (9.1)$$

$$= E\left[\sum_i I(\tau_n = s_{n,i})(Y_n(s_{n,i} + \delta_n) - Y_n(s_{n,i}))^2\right]$$

$$= E\left[\sum_i I(\tau_n = s_{n,i})E[(Y_n(s_{n,i} + \delta_n) - Y_n(s_{n,i}))^2 \mid \mathcal{H}_n(s_{n,i})]\right].$$

Now,

$$E[(Y_n(s_{n,i} + \delta_n) - Y_n(s_{n,i}))^2 \mid \mathcal{H}_n(s_{n,i})] \quad (9.2)$$
$$= E[(X_n(f(s_{n,i} + \delta_n)) - X_n(f(s_{n,i})))^2 \mid \mathcal{F}_n(f(s_{n,i}))]$$
$$= E[E[(X_n(f(s_{n,i} + \delta_n) \setminus f(s_{n,i})))^2$$
$$\mid \mathcal{G}_n^*(f(s_{n,i} + \delta_n) \setminus f(s_{n,i}))] \mid \mathcal{F}_n(f(s_{n,i}))]$$
$$= E[E[Q_n(f(s_{n,i} + \delta_n) \setminus f(s_{n,i}))$$
$$\mid \mathcal{G}_n^*(f(s_{n,i} + \delta_n) \setminus f(s_{n,i}))] \mid \mathcal{F}_n(f(s_{n,i}))]$$
$$= E[(\lambda_n(s_{n,i} + \delta_n) - \lambda_n(s_{n,i})) \mid \mathcal{H}_n(s_{n,i})].$$

Substituting (9.2) in (9.1),

$$E[(Y_n(\tau_n + \delta_n) - Y_n(\tau_n))^2] \quad (9.3)$$
$$= E\left[\sum_i I(\tau_n = s_{n,i}) E[(\lambda_n(s_{n,i} + \delta_n) - \lambda_n(s_{n,i})) \mid \mathcal{H}_n(s_{n,i})]\right]$$
$$= E[\lambda_n(\tau_n + \delta_n) - \lambda_n(\tau_n)].$$

Now, we have that $\lambda_n(s) \to_P \lambda(s)$, $\forall s \in [0,1]$, and since λ_n is increasing $\forall n$ and λ is continuous and deterministic, as observed by McLeish (1978), it follows that $\sup_{0 \leq s \leq 1} \mid \lambda_n(s) - \lambda(s) \mid \to_P 0$. Therefore, given $\epsilon > 0$, for all n sufficiently large that

$$\sup_{0 \leq s \leq 1-\delta_n} \mid \lambda(s + \delta_n) - \lambda(s) \mid < \epsilon/3,$$

$$P(\mid \lambda_n(\tau_n + \delta_n) - \lambda(\tau_n) \mid > \epsilon)$$
$$\leq 2P(\sup_{0 \leq s \leq 1} \mid \lambda_n(s) - \lambda(s) \mid > \epsilon/3)$$
$$\to_P 0.$$

Finally, since $\lambda_n(\tau_n + \delta_n) - \lambda(\tau_n) < \lambda_n(1) = Q_n(T)$, by uniform integrability we have

$$E[(Y_n(\tau_n + \delta_n) - Y_n(\tau_n))^2] = E[\lambda_n(\tau_n + \delta_n) - \lambda(\tau_n)] \to 0.$$

Thus, condition (ii) of Theorem 8.2.1 has been proven, and the sequence (Y_n) is tight in $D[0,1]$.

Proof of (II):
Let $Y : (\Omega', \mathcal{F}', P') \to D[0,1]$ be any weak limit of a subsequence $(Y_{n'})$ of (Y_n), where $(\Omega', \mathcal{F}', P')$ is assumed to be complete. We recall that $E[|Y_n(1)|^{2+\delta}]$ is uniformly bounded, ensuring that Y is the weak limit of a sequence $(Y_{n'})$ of martingales such that $(Y_{n'}(s))_{n',s}$ is uniformly integrable, and as observed above, Y is continuous

CENTRAL LIMIT THEOREMS

with probability 1. That Y is a martingale is well known (see, for example, Jacod and Shiryaev (1987), Proposition IX.1.12), so we need only show that $\lambda = \Lambda \circ f$ is a quadratic variation for Y with respect to its minimal filtration $\mathcal{H}^0$: i.e., that for any $0 \leq u \leq v \leq 1$,

$$E[(Y(v) - Y(u))^2 \mid \mathcal{H}^0(u)] = E[(\lambda(v) - \lambda(u)) \mid \mathcal{H}^0(u)].$$

By Dynkin's $\pi - \lambda$ theorem, it is enough to show that for any $d \in \mathbf{N}$, $0 \leq s_1 \leq \ldots \leq s_d \leq u \leq v \leq 1$, and $(r_1, \ldots, r_d) \in \mathbf{R}^d$ such that r_i is a P'-continuity point of $Y(s_i)$, $i = 1, \ldots, d$,

$$E[I(Y(s_1) \leq r_1, \ldots, Y(s_d) \leq r_d)(Y(v) - Y(u))^2] \quad (9.4)$$
$$= E[I(Y(s_1) \leq r_1, \ldots, Y(s_d) \leq r_d)(\lambda(v) - \lambda(u))].$$

We note that for each n, as shown in (9.2) above,

$$E[(Y_n(v) - Y_n(u))^2 \mid \mathcal{H}_n(u)] = E[(\lambda_n(v) - \lambda_n(u)) \mid \mathcal{H}_n(u)].$$

In particular,

$$E[I(Y_n(s_1) \leq r_1, \ldots, Y_n(s_d) \leq r_d)(Y_n(v) - Y_n(u))^2] \quad (9.5)$$
$$= E[I(Y_n(s_1) \leq r_1, \ldots, Y_n(s_d) \leq r_d)(\lambda_n(v) - \lambda_n(u))].$$

Now, since (Y_n) converges to Y in the J_1-topology and since Y is continuous, all of the corresponding finite dimensional distributions converge. Therefore, if r and $-r$ are P'-continuity points of $Y(v) - Y(u)$,

$$P_n[(I(Y_n(s_1) \leq r_1, \ldots, Y_n(s_d) \leq r_d)(Y_n(v) - Y_n(u))^2) \leq r^2]$$
$$= P_n[(Y_n(s_1) \leq r_1, \ldots, Y_n(s_d) \leq r_d)^c]$$
$$+ P_n[Y_n(s_1) \leq r_1, \ldots, Y_n(s_d) \leq r_d, -r \leq (Y_n(v) - Y_n(u)) \leq r]$$
$$\to P'[(Y(s_1) \leq r_1, \ldots, Y(s_d) \leq r_d)^c]$$
$$+ P'[Y(s_1) \leq r_1, \ldots, Y(s_d) \leq r_d, -r \leq (Y(v) - Y(u)) \leq r]$$
$$= P'[(I(Y(s_1) \leq r_1, \ldots, Y(s_d) \leq r_d)(Y(v) - Y(u))^2) \leq r^2].$$

Thus, $(I(Y_n(s_1) \leq r_1, \ldots, Y_n(s_d) \leq r_d)(Y_n(v) - Y_n(u))^2)_n$ converges in distribution to $I(Y(s_1) \leq r_1, \ldots, Y(s_d) \leq r_d)(Y(v) - Y(u))^2$, and by the uniform bound on $E[|Y_n(1)|^{2+\delta}]$, the sequence is uniformly integrable. Likewise, since $\lambda_n(v) - \lambda_n(u) \to_P \lambda(v) - \lambda(u)$, $(I(Y_n(s_1) \leq r_1, \ldots, Y_n(s_d) \leq r_d)(\lambda_n(v) - \lambda_n(u)))_n$ converges in distribution to $I(Y(s_1) \leq r_1, \ldots, Y(s_d) \leq r_d)(\lambda(v) - \lambda(u))$, and this sequence is also uniformly integrable. Hence, we may take limits in (9.5) to prove (9.4). This completes the proof of (II).

To summarize, we have shown that any weak limit Y of $(Y_n) = (M_f(X_n))$ is a martingale which is almost surely continuous, and which has quadratic variation $\Lambda \circ f$. Thus, since f was arbitrary, there exists a Brownian motion X with variance function Λ such that $X_n \to_{sf} X$. □

We recall from Section 7.3 that semi-functional convergence is appropriate for any limiting Brownian motion process, regardless of whether or not an outer-continous version (on $\mathcal{A}$) exists. However, in the event that the class $\mathcal{A}$ admits a continuous version of the limiting Brownian motion, we can prove a functional central limit theorem. First, we recall some notation and definitions from Section 7.1. $\mathcal{D}(\mathcal{A})$ is the class of functions in $B(\mathcal{A})$ which are outer-continuous with inner limits, and $C(\mathcal{A})$ is the class of functions which are continuous with respect to the Hausdorff metric on $\mathcal{A}$. The class

$$\Xi\left(\boldsymbol{\Gamma},\boldsymbol{\Sigma},\boldsymbol{\Delta}\right) = \Xi\left((\Gamma_n)_n, (\Sigma_n)_n, (\Delta_n)_n\right)$$

is defined in Definition 7.1.10, and consists of functions x which can be uniformly approximated by a sequence of sums of the form $J_n(x) + C_n(x)$, where $J_n(x) \in \Gamma_n$ is purely atomic and $C_n(x) \in \Sigma_n$ is continuous. It was proven in Theorem 7.1.15 that $\Xi\left(\boldsymbol{\Gamma},\boldsymbol{\Sigma},\boldsymbol{\Delta}\right)$ is compact whenever both (Γ_n) and (Σ_n) are sequences of compact sets, provided that $\mathcal{A}$ is consistent with boundaries (cf. Definition 7.1.11).

The following functional central limit theorem is now a corollary of Theorem 9.1.2. A version of this theorem specialized to a sequence of purely atomic processes was given in Slonowsky (1998).

Theorem 9.1.4 *Assume that $\mathcal{A}$ is consistent with boundaries, that Λ is a deterministic, increasing, monotone inner- and outer-continuous function and that there exists a Brownian motion process $X \in C(\mathcal{A})$ with variance function Λ. Let $(X_n, \mathcal{F}_n, P_n)$ be a sequence of strong martingales defined on $\mathcal{A}$, and let (Q_n) be any sequence of corresponding *-quadratic variation processes. Assume that*

1. *$\sup_n E[|\,X_n(T)\,|^{2+\delta}] < \infty$.*
2. *$J(M_f(X_n)) \to_P 0$ as $n \to \infty$ for every simple flow f.*
3. *$\{Q_n(T)\}$ is uniformly integrable.*
4. *For every $A \in \mathcal{A}$, $Q_n(A) \to_P \Lambda(A)$.*
5. *For every $\epsilon > 0$ there exists a triple $(\boldsymbol{\Gamma}_\epsilon, \boldsymbol{\Sigma}_\epsilon, \boldsymbol{\Delta}_\epsilon)$ as in Definition*

7.1.10, satisfying the conditions of Theorem 7.1.15, such that for every n,
$$P[X_n \in \Xi(\Gamma_\epsilon, \Sigma_\epsilon, \Delta_\epsilon)] \geq 1 - \epsilon.$$
Then
$$X_n \to_{\mathcal{D}} X.$$

Proof. By Theorem 9.1.2, conditions 1-4 imply that $X_n \to_{sf} X$. Since $X \in C(\mathcal{A})$, by Proposition 7.3.7 it follows that $X_n \to_{fdd} X$. Therefore, by Theorem 7.2.6, condition 5 then implies that $X_n \to_{\mathcal{D}} X$. □

9.2 The Weighted Empirical Process

In this section, we shall show that a weighted empirical process satisfies the conditions of Theorem 9.1.2. This example of a strong martingale was suggested to us by Burke (1997) and appears in Slonowsky and Ivanoff (1999).

For simplicity, we shall assume that $T = [0,1]^d$ and that $\mathcal{A}$ is the class of lower sets in T. Let F be a distribution function on $[0,1]^d$ and let $(Y_1, Y_2, ...)$ be a sequence of independent, identically distributed T-valued random vectors with distribution F. We shall also view F as a measure on the Borel sets in T, and we shall assume that F is absolutely continuous with respect to Lebesgue measure. Next, let $(Z_1, Z_2, ...)$ be a sequence of i.i.d. random variables, independent of (Y_n) with $P(Z_1 = 0) = 0$, $E(Z_1) = 0$ and $\text{Var}(Z_1) = 1$. We shall assume that the underlying probability space $(\Omega, \mathcal{F}, P)$ is complete.

We define the nth weighted empirical process $U_n = \{U_n(A) : A \in \mathcal{A}\}$ by:
$$U_n(A) = n^{-1/2} \sum_{i=1}^{n} I_{\{Y_i \in A\}} Z_i.$$

In the case in which $\mathcal{A} = \{[0,t] : t \in T\}$ and $Z_i \sim N(0,1)$, Burke (1997) has shown weak convergence of (U_n) as a sequence of multiparameter processes in the Skorokhod J_1-topology on $D([0,1]^d)$. We shall prove a semi-functional CLT on the class of lower sets, where, as pointed out previously, we cannot prove a functional CLT since the limiting Brownian motion does not have sample paths in $\mathcal{D}(\mathcal{A})$.

It is an immediate consequence of absolute continuity that F

can be regarded as an inner- and outer-continuous increasing (deterministic) process on $\mathcal{A}$. This will be the variance function of the limiting Brownian motion.

We will now show that U_n is a strong martingale and we will find an appropriate *-quadratic variation.

U_n is a strong martingale:

We must first define a suitable filtration. If $W_i(A) = I_{\{Y_i \in A\}} Z_i$, then $n^{1/2} U_n(A) = \sum_{i=1}^n W_i(A)$, and (W_i) is an i.i.d. sequence of $\mathcal{A}$-indexed processes. Define

$$\mathcal{F}_A = \cap_m \mathcal{F}^0_{g_m(A)},$$

where $\mathcal{F}^0_{g_m(A)} = \sigma(W_i(A') : i \in \mathbf{N}, A' \subseteq g_m(A), A' \in \mathcal{A}) \vee \mathcal{F}_0$, and $\mathcal{F}_0$ consists of all the P-null subsets of Ω. Clearly $(\mathcal{F}_A)_{A \in \mathcal{A}}$ is complete and outer-continuous, and for $A \setminus B = C \in \mathcal{C}$ a maximal representation of C in T, $\mathcal{G}^*_C = \mathcal{F}_B$. If it can be shown that W_i is a strong martingale with respect to $\mathcal{F}$, then it will follow that U_n is as well.

Trivially W_i is adapted, and for $A \setminus B = C \in \mathcal{C}$ a maximal representation of C in T,

$$
\begin{aligned}
& E[W_i(C) \mid \mathcal{G}^*_C] \\
&= E[W_i(C) \mid \mathcal{F}_B] \\
&= \lim_m E[W_i(C) \mid \mathcal{F}^0_{g_m(B)}] \qquad (9.6) \\
&= \lim_m \left(E[W_i(A \setminus g_m(B)) \mid \mathcal{F}^0_{g_m(B)}] \right) \\
&\quad + \lim_m (W_i(g_m(B) \setminus B)) \\
&= \lim_m \left(E[I_{\{Y_i \in A \setminus g_m(B)\}} Z_i \mid \mathcal{F}^0_{g_m(B)}] \right) \\
&\quad + \lim_m (W_i(g_m(B) \setminus B)) \\
&= \lim_m \left(I_{\{Y_i \notin g_m(B)\}} E[Z_i] \frac{F(A \setminus g_m(B))}{(1 - F(g_m(B)))} \right) \qquad (9.7) \\
&\quad + \lim_m (W_i(g_m(B) \setminus B)) \\
&= I_{\{Y_i \notin B\}} E[Z_i] \frac{F(C)}{(1 - F(B))} + 0 \qquad (9.8) \\
&= 0. \qquad (9.9)
\end{aligned}
$$

(9.6) follows by the reverse martingale convergence theorem.

THE WEIGHTED EMPIRICAL PROCESS

(9.7) holds by completeness of the σ-algebras, since

$$\{W_i(g_m(B)) = 0\} \setminus \{Y_i \notin g_m(B)\} \subseteq \{Z_i = 0\}.$$

As $P(Z_i = 0) = 0$, $\{Y_i \notin g_m(B)\} \in \mathcal{F}^0_{g_m(B)}$. As well, Y_i and Z_i are independent of each other and of $(W_j)_{j \neq i}$.

Thus, each W_i is a strong martingale with respect to $\mathcal{F}$, and so U_n is as well.

A *-quadratic variation for U_n:

As above, let $A \setminus B = C \in \mathcal{C}$ be a maximal representation of C in T. We observe that

$$n(U_n(C))^2 = \sum_{i=1}^{n}(W_i(C))^2 + \sum_{i \neq j} W_i(C)W_j(C).$$

By independence of W_i and W_j, it is easily seen that

$$\begin{aligned} E[W_i(C)W_j(C) \mid \mathcal{G}_C^*] &= E[W_i(C) \mid \mathcal{G}_C^*]E[W_j(C) \mid \mathcal{G}_C^*] \\ &= 0 \text{ by (9.9).} \end{aligned} \quad (9.10)$$

Next, arguing as in (9.8),

$$\begin{aligned} E[(W_i(C))^2 \mid \mathcal{G}_C^*] &= E[I_{\{Y_i \in C\}} Z_i^2 \mid \mathcal{G}_C^*] \\ &= I_{\{Y_i \notin B\}} E[Z_i^2] \frac{F(C)}{(1 - F(B))} \\ &= I_{\{Y_i \notin B\}} \frac{F(C)}{(1 - F(B))}, \end{aligned} \quad (9.11)$$

where, as observed before, $\{Y_i \notin B\} \in \mathcal{F}_B = \mathcal{G}_C^*$. Therefore,

$$E[n(U_n(C))^2 \mid \mathcal{G}_C^*] = \sum_{i=1}^{n} I_{\{Y_i \notin B\}} \frac{F(C)}{(1 - F(B))}. \quad (9.12)$$

We now recall the following notation (cf. Section 4.5): for $t = (t_1, ..., t_d)$, $R_t = \prod_{i=1}^{d}[0, t_i]$ and $S_t = \prod_{i=1}^{d}(t_i, 1]$. Define Q_n as follows:

$$Q_n(A) = n^{-1} \sum_{i=1}^{n} \int_{A \cap R_{Y_i}} (F(\overline{S_u}))^{-1} dF(u). \quad (9.13)$$

(By convention, we set $(F(\overline{S_u}))^{-1} = 0$ if $F(\overline{S_u}) = 0$.)

Lemma 9.2.1 Q_n is a *-quadratic variation for U_n.

Proof. We proceed much as in the proof of Theorem 4.5.3. Let

$C = A \setminus B$ be a maximal representation of $C \in \mathcal{C}$. By independence of the Y_i's

$$nE[Q_n(C) \mid \mathcal{G}_C^*]$$
$$= \sum_{i=1}^n \frac{I_{\{Y_i \notin B\}}}{1 - F(B)} \int_{B^c} \int_{C \cap R_v} (F(\overline{S_u}))^{-1} dF(u) dF(v)$$
$$= \sum_{i=1}^n \frac{I_{\{Y_i \notin B\}}}{1 - F(B)} \int_C (F(\overline{S_u}))^{-1} \int_{\overline{S_u}} dF(v) dF(u) \quad (9.14)$$
$$= \sum_{i=1}^n I_{\{Y_i \notin B\}} \frac{F(C)}{(1 - F(B))}. \quad (9.15)$$

When changing the order of integration in (9.14), we note that $u \in R_v$ if and only if $v \in \overline{S_u}$.

Now, by (9.12) and (9.15), it follows that

$$E[(U_n(C))^2 \mid \mathcal{G}_C^*] = E[Q_n(C) \mid \mathcal{G}_C^*]$$

and the Lemma is proven. □

We can now proceed with the central limit theorem.

Theorem 9.2.2 *If $E[(Z_i)^4] < \infty$, as $n \to \infty$,*

$$U_n \to_{sf} X,$$

where X is a Brownian motion with variance measure F.

Proof. We must verify the conditions of Theorem 9.1.2.

1. Note that $U_n(T) = \sum_{i=1}^n Z_i$ and recall that the Z_i's are independent with mean 0 and variance 1. Therefore,

$$E[(U_n(T))^4] = n^{-2} E[(\sum_{i=1}^n Z_i)^4]$$
$$= n^{-2} \left(nE(Z_1^4) + 3n(n-1)[E(Z_1^2)]^2 \right)$$
$$\leq E(Z_1^4) + 3.$$

Therefore, condition 1 is satisfied with $\delta = 2$.

2. Consider a simple flow f, and the process $M_f(U_n)$. $M_f(U_n)$ is equal in distribution to a weighted empirical process V_n on $[0,1]$ corresponding to the distribution function $F_f(s) = F \circ f(s)$, $0 \leq s \leq 1$. Since F is absolutely continuous and $f(s)$ is a lower set, the continuity of the flow f ensures that F_f is a continuous function on

THE WEIGHTED EMPIRICAL PROCESS   185

[0, 1]. Therefore, with probability 1, $J(V_n) \leq n^{-1/2} \max_{1 \leq i \leq n} |Z_i|$, and since $E[\max_{1 \leq i \leq n} |Z_i|^2] = o(n)$,

$$J(M_f(U_n)) \to_P 0$$

as $n \to \infty$ for every simple flow f.

3. When verifying condition 4, we shall show that $Q_n(T) \to_P F(T)$ and that $E[Q_n(T)] = F(T)$ for every n. This implies that $(Q_n(T))_n$ is uniformly integrable (cf. Billingsley (1968), Theorem 5.4).

4. We shall show that $Q_n(A) \to_P F(A)$, for every set $A \in \mathcal{A}$ of the form $A = R_t$. If this is the case, then it follows that $Q_n(A) \to_P F(A)$ for $A \in \mathcal{A}$ of the form $A = \cup_{h=1}^k R_{t_h}$. For an arbitrary lower set A, the absolute continuity of F and the fact that A can be approximated uniformly by finite unions of rectangles permits us to conclude that $Q_n(A) \to_P F(A)$.

To avoid technicalities in what follows, we shall assume that the density f of F with respect to Lebesgue measure is strictly positive on T. We shall first prove that $Q_n(A) \to_{L^2} F(A)$, for every set $A = R_t$ where $t = (t_1, ..., t_d)$ is such that $t_i < 1, \forall i = 1, ..., d$. We denote this by $t \ll 1$. Note in this case that $F(S_t) > 0$. Now, it is clear from (9.15) that

$$E[Q_n(C)] = F(C) \text{ for every } C \in \mathcal{C}. \tag{9.16}$$

Therefore, if $D = R_v$ where $v_i = 1$ for at least one i, a sequence $(t_m) = (t_{m,1}, ..., t_{m,d})_m$ exists with $t_m \ll 1, \forall m$ and such that $R_{t_m} \subset D$ and $F(D \setminus R_{t_m}) < 1/m$. Hence, by (9.16)

$$P(|Q_n(D) - Q_n(R_{t_m})| \geq \epsilon) \leq 1/m\epsilon$$

for every n. Now, if $Q_n(R_{t_m}) \to_P F(R_{t_m})$ as $n \to \infty$ and since $F(R_{t_m}) \to F(D)$ as $m \to \infty$, it follows that $Q_n(D) \to_P F(D)$ as $n \to \infty$. Hence, it suffices to show that for $t \ll 1$ and $A = R_t$,

$$E[(Q_n(A))^2] \to (F(A))^2. \tag{9.17}$$

$$Q_n(A)^2 = n^{-2} \sum_{i=1}^n \left[\int_{R_{Y_i} \cap A} (F(S_u))^{-1} dF(u) \right]^2 \tag{9.18}$$

$$+ n^{-2} \sum_{i \neq j} \int_{R_{Y_i} \cap A} (F(S_u))^{-1} dF(u)$$

$$\cdot \int_{R_{Y_j} \cap A} (F(S_v))^{-1} dF(v), \tag{9.19}$$

where, by absolute continuity, we have replaced $\overline{S_u}$ with S_u.

By independence, the expected value of (9.19) is

$$n^{-2}n(n-1)(F(A))^2 \to (F(A))^2.$$

Thus, (9.17) will follow if it can be shown that the expected value of (9.18) converges to 0. As observed in the discussion of the empirical process in Chapter 8, we may apply a d-dimensional integration by parts formula to show that for an integrable function $f : T \to \mathbf{R}$, $(f(t))^2$ may be expressed as a finite linear combination of integrals of the form $\int_{R_t} f(u')df(u)$, where u' is such that $u'_j = u_j \; \forall j \in J$ where J is some nonempty subset of $\{1, ..., d\}$, and $u'_j = t_j, \; \forall j \notin J$. If we let

$$f(t) = \int_{R_t \cap R_{Y_i}} (F(S_u))^{-1} dF(u),$$

then

$$E\left[\int_{R_t} f(u')df(u)\right]$$
$$= \int_{R_t} \int_{R_{u'}} \frac{F(S_u \cap S_v)}{F(S_u)F(S_v)} dF(v)dF(u)$$
$$\leq \left[\int_{R_t} (F(S_u))^{-1/2} dF(u)\right]^2$$
$$\leq (F(S_t))^{-1}. \tag{9.20}$$

We observe that (9.20) implies that there exists an integer K such that

$$E\left(n^{-2}\sum_{i=1}^{n}\left[\int_{R_{Y_i} \cap A} (F(S_u))^{-1} dF(u)\right]^2\right) \leq n^{-1}K(F(S_t))^{-1}$$

and so the expected value of (9.18) converges to 0, as required. This completes the proof of both conditions 3 and 4, and the central limit theorem follows. □

References

Adler, R. J., *The Geometry of Random Fields*, Wiley, Chichester, 1981.

Adler, R. J., *An introduction to continuity, extrema and related topics for general Gaussian processes*, IMS Lect. Notes - Monograph series, Vol. 12, Institute of Mathematical Statistics, Hayward, California, 1990.

Adler, R. J., and Pyke R., Uniform quadratic variations for Gaussian processes, *Stoch. Proc. and their Appl.* 48 (1993), 191-210.

Adler, R. J., Monrad, D., Scissors, R. and Wilson, R. J., Representations, decompositions, and sample function continuity of random fields with independent increments, *Stoch. Proc. and their Appl.* 15 (1983), 3-30.

Alabert, A., and Nualart, D., Some remarks on the conditional independence and the Markov property. *Stoch. Anal. Rel. Topics*, H. Korezlioglu, A.S. Ustunel (eds.), Probability Series 31(1992), Birkhauser, Boston, 343-364.

Aldous, D., Stopping times and tightness, *Ann. Probab.* 6 (1978), 335-340.

Aldous, D., Stopping times and tightness II, *Ann. Probab.* 17 (1989), 586-595.

Alexander, K. S., Sample moduli for set-indexed Gaussian processes, *Ann. Probab.* 14 (1986), 598-611.

Alexander, K. S., and Pyke, R., A uniform central limit theorem for set-indexed partial sum processes with finite variance, *Ann. Probab.* 14 (1986), 582-597.

Al Hussaini, A., and Elliott, R. J., Martingales, potentials and exponentials associated with a two-parameter process, *Stochastics* 6 (1981), 23-42.

Al Hussaini, A., and Elliott, R. J., Convergence of the empirical distribution to the Poisson process, *Stochastics* 13 (1984), 299-308.

Al Hussaini, A., and Elliott, R. J., Filtrations for the two-parameter jump process, *J. Multiv. Anal.* 16 (1985), 118-139.

Allain, M. F., Tribus prévisibles et espaces de processus à trajectoires continues indexés par un espace localement compact et métrisable, *Sém. de Rennes*, 1979.

Allain, M. F., Mesures stochastiques et décomposition de Doob, *Sém. de Rennes*, 1983.

Allain, M. F., Semimartingales indexées par une partie R^d et formule d'Ito, cas continu., *Z. Wahrsch. Verw. Gebiete* 65 (1984a), 421-444.

Allain, M. F., Caractérisation de mesures stochastiques à valeurs dans L_0, *Sém. de Rennes*, 1984b.

Astbury, K. A., Amarts indexed by directed sets, *Ann. Probab.* 6 (1978), 267-278.

Astbury, K. A., The order convergence of martingales indexed by directed sets, *Trans. Amer. Math. Soc.* 265 (1981), 495-510.

Baigger, G., Summability of martingales with two dimensional parameter, *Math. Z.* 178 (1981), 381-386.

Bakry, D., Sur la régularité des trajectoires des martingales à deux indices, *Z. Wahrsch. verw. Gebiete* 50 (1979), 149-157.

Bakry, D., Limites quadrantales des martingales, Colloque ENST-CNET, *Lect. Notes in Math.* 863 (1981), 40-49.

Bakry, D., Semimartingales à deux indices, Sém. de Probab. XVI, *Lect. Notes in Math.* 920 (1982), 355-369.

Bass, R. F., and Pyke, R., The existence of set-indexed Lévy processes, *Z. Wahrsch. verw. Gebiete* 66 (1984a), 157-172.

Bass, R. F., and Pyke, R., Functional law of the iterated logarithm and uniform central limit theorem for partial-sum processes indexed by sets, *Ann. Probab.* 12 (1984b), 13-34.

Bass, R. F., and Pyke, R., A strong law of large number for partial-sum processes indexed by sets, *Ann. Probab.* 12 (1984c), 268-271.

Bass, R. F., and Pyke, R., The space $D(A)$ and weak convergence for set-indexed processes, *Ann. Probab.* 13 (1985), 860-884.

Bass, R. F., and Pyke, R., A central limit theorem for $D(A)$-valued processes, *Stoch. Proc. and their Applic.* 24 (1987), 109-131.

Belyaev, Y. K., Continuity and Holder's conditions for sample functions of stationary Gaussian processes, *Proc. Fourth Berkeley Symp. on Math. Stat. and Probability* 2 (1961), 23-33.

Berman, S. M., Gaussian processes with stationary increments: Local times and sample function properties, *Ann. Math. Stat.* 41 (1970), 1260-1272.

Bernard, P., Espaces H_1 de martingales à deux indices, dualité avec les martingales de type 'BMO', *Bull. Sci. Math.* 103 (1979), 297-303.

Bickel, P. J., and Wichura, M. J., Convergence criteria for multi-parameter stochastic processes and some applications, *Ann. Math. Stat.* 42 (1971), 1656-1670.

Billingsley, P. *Convergence of Probability Measures*, John Wiley, New York, 1968.

Blei, R. C., Multilinear measure theory and applications to stochastic integration, *Probab. Th. Rel. Fields* 81 (1989), 569-584.

Blei, R. C., Stochastic integrators indexed by a multi-dimensional parameter, *Probab. Th. Rel. Fields* 95 (1993), 141-154.

Bochner, S., Partial-ordering in the theory of martingales, *Ann. Math.* 62 (1955), 162-169.

Boucher, C., Ellis, R., and Turkington, B., Spatializating random measures: Doubly indexed processes and the large deviation principle, *Ann. Probab.* 27 (1999), 297-324.

Bouleau, N., Sur la variation quadratique de certaines mesures vectorielles, *Z. Wahrsch. verw. Gebiete* 61 (1982), 283-290.

Brémaud, P., *Point Processes and Queues*, Springer Series in Statistics, Springer-Verlag, New York, 1981.

Brennan, M. D., Planar semimartingales, *J. Multiv. Anal.* 9 (1979), 465-486.

Brossard, J., Comparaison des normes L_p du processus croissant et de la variable maximale pour les martingales régulières à deux indices, théorème local correspondant, *Ann. Probab.* 8 (1980), 1183-1188.

Brossard, J., Regularité des martingales à deux indices et inégalités de normes, Colloque ENST-CNET, *Lect. Notes in Math.* 863 (1981), 91-121.

Brossard, J., and Chevalier, L., Calcul stochastique et inégalités de normes pour les martingales bi-browniennes, Applications aux fonctions bi-harmoniques, *Ann. Inst. Fourier* 30 (1980), 97-120.

Brown, T., A martingale approach to the Poisson convergence of simple point processes, *Ann. Probab.* 6 (1978), 615-628.

Brown, T., Ivanoff, G., and Weber, N. C., Poisson convergence in two dimensions with applications to row and column exchangeable arrays, *Stoch. Proc. and their Applic.* 23 (1986), 307-318.

Burke, M. D., Multivariate test-of-fit and uniform confidence bands using a Gaussian bootstrap, *Unpublished manuscript* (1997).

Burstein, L., Generalized set-indexed martingales, Ph. D. Thesis, Bar-Ilan University, 1999.

Cabaña, E. M., On a martingale characterization of two-parameter Wiener process, *Stat. Probab. Letters* 10 (1990), 263-270.

Cabaña, E. M., and Wschebor, M., Sur le processus de Wiener à deux paramètres, *C.R. Acad. Sci. Paris* 289 (1979), 453-455.

Cabaña, E. M., and Wschebor, M., An estimate for the tails of the distribution of the supremum for a class of stationary multiparameter Gaussian processes, *J. Appl. Probab.* 18 (1981), 536-541.

Cabaña, E. M., and Wschebor, M., The two-parameter Brownian bridge: Kolmogorov inequalities and upper and lower bounds for the distribution of the maximum, *Ann. Probab.* 10 (1982), 289-302.

Cairoli, R., Sur la convergence des martingales indexées par $N \times N$, Sém. de Probab. XIII, *Lect. Notes in Math.* 721 (1979), 162-173.

Cairoli, R., and Dalang, R.C., *Sequential Stochastic Optimization*, J.Wiley, New York, 1996.

Cairoli, R., and Gabriel, J. P., Arrêt de certaines suites multiples de variables aléatoires indépendantes, Sém. de Probab. XIII, *Lect. Notes in Math.* 721 (1979), 174-198.

Cairoli, R., and Walsh, J. B., Stochastic integrals in the plane, *Acta Math.* 134 (1975), 111-183.

Cairoli, R., and Walsh, J. B., On changing time, Sém. de Probab. XI, *Lect. Notes in Math.* 581 (1977), 349-355.

Cairoli, R., and Walsh, J. B., Régions d'arrêt, localisations et prolongements de martingales, *Z. Wahrsch. verw. Gebiete* 44 (1978), 279-306.

Capasso, V., De Giosa, M., and Mininni, R., Characterization of the spatial Poisson process by stopping lines, *Stochastics and Stochastics Rep.*, to appear.

Cartier, P., Introduction à l'étude des mouvements Browniens à plusieurs paramètres, Sém. de Probab. V, *Lect. Notes in Math.* 191 (1971), 58-75.

Chentsov, N. N., Lévy Brownian motion for several parameters and generalized white noise, *Th. Probab. and its Appl.* 2 (1957), 665-766.

Chevalier, L., L_p inequalities for two-parameter martingales, Stochastic Integrals, *Lect. Notes in Math.* 851 (1981), 470-475.

Chow, Y. S., Martingales in a σ-finite measure space indexed by directed sets, *Trans. A.M.S.* 97(1960), 254-285.

Chow, Y. S., A martingale convergence theorem of Ward's type, *Illinois J. Math.* 9 (1965), 569-576.

Christofides, T. C., and Serfling, R. J., Maximal inequalities for multidimensionally indexed submartingales arrays, *Ann. Probab.* 18 (1990), 630-641.

Clarkson, J. A., and Adams, C. R., On definition of bounded variation for functions of two variables, *Trans. A.M.S.* 35 (1933), 824-854.

Cojocaru, L., and Merzbach, E., Strictly simple set-indexed point processes and Cox processes, preprint, 1999.

Csörgo, M., and Revesz, P., How big are the increments of a multiparameter Wiener process? *Z. Wahrsch. verw. Gebiete* 42 (1978), 1-12.

Dabrowska, D. M., Kaplan-Meier estimate on the plane, *Ann. of Statist.* 16 (1988), 1475-1489.

Dabrowski, A. R., Fang, Y. and Ivanoff, B. G., Strong approximations for multiparameter martingales, *Stochastics and Stochastics Rep.* 56 (1996), 241-270.

Dalang, R. C., Sur l'arrêt optimal de processus à temps multidimensionnel continu. Sém. de Probab. XVIII, *Lect. Notes in Math.* 1059 (1984), 379-390.

Dalang, R. C., On infinite perfect graphs and randomized stopping points on the plane, *Probab. Th. Rel. Fields* 78 (1988), 357-378.

Dalang, R. C., Randomization in the two-armed bandit problem, *Ann. Probab.* 18 (1990), 218-225.

Dalang, R. C., and Walsh, J. B., The sharp Markov property for Lévy sheets, *Ann. Probab.* 20 (1992), 591-626.

Dalang, R. C., and Walsh, J. B., The sharp Markov property of the Brownian sheet and related processes, *Acta Math.* 168 (1992), 153-218.

Daley, D. J., and Vere-Jones, D., *An Introduction to the Theory of Point Processes*, Springer Series in Statistics, Springer-Verlag, New York, 1988.

Davidsen, M. and Jacobsen, M., Weak convergence of two-sided stochastic integrals, with an application to models for left truncated survival data, *Statistical Inference in Stochastic Processes*, Prabhu, N.U. and Basawa, I.V. (eds.), 167-182, M. Dekker, 1991.

Davydov, Y. A., Local times for multiparameter random processes, *Th. Probab. Appl.* 23 (1978), 573-583.

De Giosa, M., and Mininni, R., On the Doléans function of set-indexed submartingales, *Stat. Probab. Letters* 24 (1995), 71-75.

Dellacherie, C., and Meyer, P. A., *Probabilités et Potentiel*, Hermann, Paris, 1980.

Dieudonné, J., Sur un théorème de Jessen, *Fund. Math.* 37 (1950), 242-248.

Dobrushin, R. L., and Surgailis, D., On the innovation problem for Gaussian Markov random fields, *Z. Wahr. verw. Gebiete* 49 (1979), 275-291.

Doss, H., and Dozzi, M., Estimations de grandes déviations pour les processus de diffusion à paramètre multidimensionnel, Sém. de Probab. XX, *Lect. Notes in Math.* 1204 (1986), 68-80.

Dozzi, M., On the decomposition and integration of two-parameter stochastic processes, Colloque ENST-CNET, *Lect. Notes in Math.* 863 (1981), 162-171.

Dozzi, M., *Stochastic Processes with a Multidimensional Parameter*, Pitman Research Notes in Math. Series 194, Longman, New York, 1989.

Dozzi, M., Ivanoff, B. G., and Merzbach, E., Doob-Meyer decomposition for set-indexed submartingales, *J. of Theor. Probab.*, 7 (1994), 499-525.

Duc, N. M., and Nguyen, X. L., On the transformation of a martingale with a two-dimensional parameter set by convex functions, *Z. Wahrsch. verw. Gebiete* 66 (1984), 19-24.

Duc, N. M., Nualart, D., and Sanz, M., Planar semimartingales obtained by transformations of two-parameter martingales, Sém. de Probab. XXIII, *Lect. Notes in Math.* 1372 (1989), 567-582.

Duc, N. M., Nualart, D., and Sanz, M., The Doob-Meyer decomposition for anticipating processes, *Stochastics and Stochastics Reports* 34 (1991), 221-239.

Dudley, R. M., Gaussian processes on several parameters, *Ann. Math.*

Stat. 36 (1965), 771-788.

Dudley, R. M., Sample functions of the Gaussian process, *Ann. Probab.* 1 (1973), 66-103.

Dudley, R. M., Lower layers in R^2 and convex sets in R^3 are not GB classes, *Lect. Notes in Math.* 709 (1979), 97-102.

Dudley, R. M., *A Course on Empirical Processes.* Lect. Notes Math., vol. 1097, 1-142, 1984.

Dynkin, E. B., Markov processes and random fields, *Bull. Am. Math. Soc.* 3 (1980), 975-1000.

Dynkin, E. B., and Mandelbaum, A., Symmetric statistics, Poisson point processes and multiple Wiener integrals, *Ann. Math. Stat.* 11 (1983), 739-745.

Edgar, G. A., and Sucheston, L., *Stopping Times and Directed Processes*, Cambridge University Press, 1992.

Ehm, W., Sample function properties of multi-parameter stable processes, *Z. Wahrsch. verw. Gebiete* 55 (1981), 195-228.

Etemadi, N., and Wang, A. T., Quadratic variation of functionals of the two-parameter Wiener process, *J. Multiv. Anal.* 6 (1976), 630-643.

Evans, S. N., Continuity properties of Gaussian stochastic processes indexed by a local field, *Proc. London Math. Soc.* 56 (1988), 380-416.

Evans, S. N., Sample path properties of Gaussian stochastic processes indexed by a local field, *Proc. London Math. Soc.* 56 (1988), 580-624.

Evstigneev, I. V., Markov times for random fields, *Th. Probab. and Applic.* 22 (1977), 563-569.

Fisher, L., A survey of the mathematical theory of multidimensional-point processes, *Stochastic Point Processes, Statistical Analysis, Theory and Applications*, P.A.W. Lewis (ed.), Wiley, New York (1972), 468-513.

Follmer, H., Quasimartingales à deux indices, *C.R. Acad. Sci. Paris* (A) 288 (1979), 61-63.

Follmer, H., Almost sure convergence of multiparameter martingales for Markov random fields, *Ann. Probab.* 12 (1984), 133-140.

Fouque, J. P., The past of a stopping point and stopping for two-parameter processes, *J. Multiv. Anal.* 13 (1983), 561-577.

Fouque, J. P., and Millet, A., Régularité à gauche des martingales fortes à plusieurs indices, *C.R. Acad. Sci. Paris* 290 (1980), 773-776.

Frangos, N., and Imkeller, P., Quadratic variation for a class of $Llog^+L$-bounded two-parameter martingales, *Ann. Probab.* 15 (1987), 1097-1111.

Frangos, N. and Imkeller, P., The continuity of the quadratic variation of two-parameter martingales, *Stoch. Proc. and their Appl.* 29 (1988), 267-279.

Frangos, N., and Sucheston, L., On covering conditions and convergence, *Lecture Notes in Math.*, 1153 (1985), 198-225.

Frangos, N., and Sucheston, L., On multiparameter ergodic and martingale theorems in infinite measure space, *Probab. Th. Rel. Fields* 71 (1986), 477-490.

Gabriel, J. P., Martingales with a countable filtering index set, *Ann. Probab.* 5 (1977), 888-898.

Geman, D., and Zinn, J., On the increments of multidimensional random fields, *Ann. Probab.* 6 (1978), 151-158.

Gierz, G., Hofmann, K. H., Keimel, K., Lawson, J.D., Mislove, M., and Scott, D. S., *A Compendium of Continuous Lattices*, Springer-Verlag, New York, 1980.

Gikhman, I. I., Square integrable difference martingales of two parameters, *Th. Probab. and Math. Stat.* 15 (1978), 21-29.

Gikhman, I. I., Two-parameter martingales, it Russian Math. Surveys 37 (1982), 1-29.

Giné, E., and Zinn, J., The law of large number for partial sum processes indexed by sets. *Ann. Probab.* 15 (1987), 154-168.

Giné, E., Hahn, M. C., and Zinn, J., Limit theorems for random sets: An application of probability in Banach space results, *Lect. Notes in Math.* 999 (1983), 112-135.

Goldie, C. M., and Greenwood, P. E., Characterizations of set-indexed Brownian motion and associated conditions for finite-dimensional convergence, *Ann. Probab.* 14 (1986), 802-816.

Goldie, C. M., and Greenwood, P. E., Variance of set-indexed sums of mixing random variables and weak convergence of set-indexed processes, *Ann. Probab.* 14 (1986), 817-839.

Grandell, J., Doubly Stochastic Point Processes, *Lect. Notes in Math.* 529, Springer-Verlag, Berlin, New York, 1976.

Grandell, J., Point processes and random measures, *Adv. Appl. Probab.* 9 (1977), 502-526.

Greenwood, P., and Evstigneev, I., A Markov evolving random field and splitting random elements, *Th. Probab. Applic.* 37 (1993), 40-42.

Gundy, R., Local convergence of a class of martingales in multidimensional time, *Ann. Probab.* 8 (1980), 607-614.

Gushchin, A. A., On the general theory of random fields on the plane, *Russian Math. Surveys* 37 (6) (1982), 55-80.

Gushchin, A. A., On absolute continuity and singularity of the distributions of random fields, *Math. U.S.S.R. Sbornik* 46 (1983), 161-170.

Gushchin, A. A., and Mishura, Y. S., The Davis inequalities and the Gundy decomposition for two-parameter strong martingales, *Theor. Probab. Math. Statistics* 44 (1992), 45-51.

Gut, A., Convergence of reverse martingales with multidimensional indices, *Duke Math. J.* 43 (1976), 269-275.

Gut, A., Convergence rates for probabilities of moderate variations for sums of random variables with multidimensional indices, *Ann. Probab.*

8 (1980), 298-313.

Guyon, X., and Prum, B., Différents types de variations produits pour une semi-martingale représentable à deux paramètres, *Ann. Inst. Fourier, Grenoble* 29 (1979), 295-317.

Guyon, X., and Prum, B., Identification et estimation de semi-martingales représentables par rapport à un brownien à un indice double, *Colloque ENST-CNET, Lect. Notes in Math.* 863 (1981), 211-232.

Guyon, X., and Prum, B. Variation-produit et formule de Itô pour les semi-martingales représentables à deux paramètres, *Z. Wahrsch. verw. Gebiete* 56 (3) (1981), 361-397.

Hajek, B., and Wong, E., Set parametered martingales and multiple stochastic integration, *Stochastic Integrals, Lect. Notes in Math.* 851 (1981), 119-151.

Hajek, B., and Wong, E., Multiple stochastic integrals: Projection and iteration, *Z. Wahrsch. verw. Gebiete* 63 (1983), 349-368.

Hall, P., *The Theory of Coverage Processes*, John Wiley, New York, 1988.

Harenbrock, M., and Schmitz, N., Optional sampling of submartingales with scanned index sets, *J. Theor. Probab.* 5, No. 2 (1992), 309-326.

Hayes, C. A., and Pauc, C.Y., *Derivation and Martingales*, Springer-Verlag, New York 1970.

Helms, L. L., Mean convergence of martingales, *Trans. A.M.S.* 87 (1958), 439-446.

Hildebrandt, T. H., *Introduction to the Theory of Integration*, Academic Press, New York, 1963.

Hürzeler, H., Quasimartingales on partially ordered sets, *J. Multiv. Anal.* 14 (1984), 34-73.

Hürzeler, H., Stochastic integration on partially ordered sets, *J. Multiv. Anal.* 17 (1985a), 279-303.

Hürzeler, H., The optional sampling theorem for processes indexed by a partially ordered set, *Ann. Probab.* 13 (1985b), 1224-1235.

Imkeller, P., Ito's formula for continuous (N,D) processes, *Z. Wahrsch. verw. Gebiete* 65 (1984), 535-562.

Imkeller, P., On changing time for two-parameter strong martingales: A counterexample, *Ann. Probab.* 14 (1986a), 1080-1084.

Imkeller, P., A note on the localization of two-parameter processes, *Probab. Th. Rel. Fields* 73 (1986b), 119-125.

Imkeller, P., Some inequalities for strong martingales, *Ann. Inst. H. Poincaré* (1986c).

Imkeller, P., Two-parameter Martingales and their Quadratic Variation, *Lect. Notes in Math.* 1308, Springer-Verlag, New York, 1988.

Imkeller, P., Regularity and integrator properties of variation processes of two-parameter martingales with jumps, *Sém de Probab. XXIII*,

Lect. Notes in Math. 1372 (1989), 536-566.

Imkeller, P., The transformation theorem for two-parameter pure jump martingales, *Probab. Th. Rel. Fields,* 89 (1991), 261-283.

Ivanoff, B. G., Stopping times and tightness in two dimensions, *Tech. Rep.1, Lab. Research in Statistics and Probab.*, Carleton Univ. and Univ. of Ottawa, Ottawa, 1983.

Ivanoff, B. G., Poisson convergence for point processes in the plane, *J. Austral. Math. Soc.* (A) 39 (1985), 253-269.

Ivanoff, B. G., Compensator approximations for point processes on the plane, *Stochastics* 28 (1989), 317-341.

Ivanoff, B. G., Stopping times and tightness for multiparameter martingales, *Stat. Probab. Letters* 28 (1996), 111-114.

Ivanoff, B.G., Lin, Y.-X., and Merzbach, E., Weak convergence of set-indexed point processes and the Poisson process, *Th. Probab. Math. Statist.* 55 (1996), 77-89.

Ivanoff, B. G., and Merzbach, E., Characterization of compensators for point processes on the plane, *Stochastics* 29 (1990a), 395-405.

Ivanoff, B. G., and Merzbach, E., Intensity-based inference for planar point processes, *J. Multiv. Analysis* 32 (1990b), 269-281.

Ivanoff, B. G., and Merzbach, E., A martingale characterization of the set-indexed Poisson process, *Stochastics and Stochastics Reports* 51 (1994), 69-82.

Ivanoff, B. G., and Merzbach, E., Stopping and set-indexed local martingales, *Stoch. Proc. and their Appl.* 57 (1995), 83-98.

Ivanoff, B. G., and Merzbach, E., A martingale characterization of the set-indexed Brownian motion, *J. Theor. Probab.* 9 (1996), 903-913.

Ivanoff, B. G., and Merzbach, E., Poisson convergence for set-indexed empirical processes, *Statistics and Probability Letters* 32 (1997), 81-86.

Ivanoff, B. G., and Merzbach, E., A Skorokhod topology for a class of set-indexed functions, in *Skorokhod's Ideas in Probability Theory*, Korolyuk, V. and Portenko, M. (eds)., to appear, 2000.

Ivanoff, B. G., Merzbach, E., and Schiopu-Kratina, I., Predictability and stopping on lattices of sets, *Probab. Th. and Rel. Fields* 97 (1993), 433-446.

Ivanoff, B. G., Merzbach, E., and Schiopu-Kratina, I., Lattices of random sets and progressivity, *Stat. and Probab. Letters* 22 (1995), 97-102.

Ivanoff, B. G., and Weber, N. C., Weak convergence of row and column exchangeable arrays, *Stochastics* 40 (1992), 1-22.

Ivanoff, B. G., and Weber, N. C., A maximal inequality and tightness for multiparameter stochastic processes, in *Asymptotic Methods in Probability and Statistics*, B. Szyszkowicz, (ed.), (1998), Elsevier Science, 359-369.

Jacod, J., Multivariate point processes: Predictable projection, Radon-Nikodym derivatives, representation of martingales, *Z. Warsch. verw.*

Gebiete 31 (1975), 235-253.

Jacod, J., Sur la convergence des processus ponctuels, *Probab. Theory and Rel. Fields* 76 (1987), 573-586.

Jacod, J., and Shiryaev, A. N., *Limit Theorems for Stochastic Processes*, Springer-Verlag, New York, 1987.

Jagers, P., Aspects of random measures and point processes, in *Advances in Probability and Related Topics* 3, P. Ney (ed.), (1974), M. Dekker, New York, 179-239.

Jagers, P., General branching processes as Markov fields, *Stoch. Proc. and their App.*, 32 (1982), 183-212.

Julia, O., and Nualart, D., The distribution of a double stochastic integral with respect to two independent Brownian sheets, *Stochastics* 25 (1988), 171-182.

Kallenberg, O., Characterization and convergence of random measures and point processes, *Z. Wahrsch. verw. Gebiete* 27 (1973), 9-21.

Kallenberg, O., On conditional intensities of point processes, *Z. Wahrsch. verw. Gebiete* 41 (1978), 205-220.

Kallenberg, O., *Random Measures,*, 4th edition, Academic Press, 1986.

Kallenberg, O., Symmetrics on random arrays and set-indexed processes, *Tech. Rep. 345, Center for stochastic processes*, University of North Carolina, Chapel Hill, North Carolina, 1991.

Kallianpur, G., and Korezlioglu, H., White noise calculus for two-parameter filtering, *Lect. Notes in Control and Info. Sci.* 96 (1986), 61-69.

Kallianpur, G., and Mandrekar, V., The Markov property for generalized Gaussian random fields, *Ann. Inst. Fourier* 24 (1974), 143-167.

Karni, S., and Merzbach, E., On the extension of bimeasures, *J. Anal. Math.* 55 (1990), 1-16.

Kendall, W. S., Contours of Brownian processes with several dimensional times, *Z. Wahrsch. verw. Gebiete* 52 (1980), 267-276.

Kent, J. T., Continuity properties for random fields, *Ann. Probab.* 17 (1989), 1432-1440.

Kopp, P. E., *Martingales and Stochastic Integrals*, Cambridge University Press, Cambridge, 1984.

Korezlioglu, H., Passage from two-parameters to infinite dimension, *Lect. Notes in Math.* 1236 (1987), 131-153.

Korezlioglu, H., Mazziotto, G. and Szpirglas, J., Non-linear filtering equations for two-parameter semimartingales, *Stoch. Proc. and their App.* 15 (1983), 239-269.

Krengel, U., and Sucheston, L., Stopping rules and tactics for processes indexed by a directed set, *J. Multiv. Anal.* 11 (1981), 199-229.

Krickeberg, K., Convergence of martingales with a directed index set, *Trans. A.M.S.* 83 (1956), 313-337.

Krickeberg, K., and Pauc, C., Martingales et dérivation, *Bull. Soc. Math.*

de France 91 (1963), 455-554.

Kuelbs, J., The invariance principle for a lattice of random variables, *Ann. Math. Stat.* 39 (1968), 382-389.

Kurtz, T. G., The optional sampling theorem for martingales indexed by directed sets, *Ann. Probab.* 8 (1980a), 675-681.

Kurtz, T. G., Representation of Markov processes as multiparameter time changes, *Ann. Probab.* 8 (1980b), 682-715.

Lacey, M. T., Limit laws for local times of the Brownian sheet, *Probab. Th. Rel. Fields*, 86 (1990), 63-85.

Last, G., Predictable projections for point process filtrations, *Probab. Th. Rel. Fields* 99 (1994), 361-388.

Lawler, G. F. and Vanderbei, R. J., Markov strategies for optimal control problems indexed by a partially ordered set, *Ann. Probab.* 11 (1983), 642-647.

Ledoux, M., Classe *LlogL* et martingales fortes à paramètre bidimensionnel, *Ann. Inst. H. Poincaré* 17 (1981), 275-280.

Ledoux, M., Transformées de Burkholder et sommabilité de martingales à deux paramètres, *Math. Z.* 181 (1982), 529-535.

Ledoux, M., Une remarque sur la convergence des martingales à deux indices, *Sém de Probab. XVII, Lect. Notes in Math.* 986 (1983), 377-383.

Ledoux, M., Arrêt par régions de $\{S_n/|n|, n \in N^2\}$, *Sém. de Probab. XVII, Lect. Notes in Math.* 986 (1983), 384-397.

Leonenko, N. N., and Mishura, Yu. S., On an invariance principle for multiparameter martingales, *Theor. Probab. and Math. Stat.* 24 (1982), 91-101.

Lévy, P., Le mouvement brownien dépendant de plusieurs paramètres, *C.R. Acad. Sci. Paris* 220 (1945), 420-422.

Lévy, P., Exemples de processus doubles de Markoff, *C.R. Acad. Sci. Paris* 226 (1948), 307-308.

Licea, G., On supermartingales with partially ordered parameter set, *Th. Probab. Appl.*, 14 (1969), 135-137.

Lindvall, T., Weak convergence of probability measures and random functions in the function space $D[0, \infty)$, *J. Appl. Probab.* 10 (1973), 109-121.

Liptser, R. S., and Shiryayev, A. N., *Statistics of Random Processes II*, Springer-Verlag, New York, 1978.

Mandelbaum, A., Continuous multi-armed bandits and multiparameter processes, *Ann. Probab.* 15, No. 4 (1987), 1527-1556.

Mandelbaum, A. and Vanderbei, R. J., Optimal stopping and supermartingales over partially ordered sets, *Z. Wahrsch. verw. Gebiete* 57 (1981), 253-264.

Mandrekar, V., Markov properties for random fields, in *Probabilistic Analysis and Related Topics* 3, A.J. Bharucha-Reid (ed.), (1983),

161- 193.

Manevitz, L., and Merzbach, E., Multi-parameter stochastic processes via non-standard analysis, *Israel Math. Conf. Proc.*, Vol. 10 (1996), 150-167.

Matheron, G., *Random Sets and Integral Geometry*, Wiley, New York, 1975.

Mazziotto, G., Two-parameter optimal stopping and bi-Markov processes, *Z. Wahrsch. verw. Gebiete* 69 (1985), 99-135.

Mazziotto, G., Two-parameter Hunt processes and a potential theory, *Ann. Probab.* 16 (1988), 600-619.

Mazziotto, G., and Merzbach, E., Regularity and decomposition of two-parameter supermartingales, *J. Multiv. Anal.* 17 (1985), 38-55.

Mazziotto, G., and Merzbach, E., Point processes indexed by directed sets, *Stoch. Proc. and their Appl.* 30 (1988), 105-119.

Mazziotto, G., and Millet., A., Stochastic control of two-parameter processes - application: the two-armed bandit problem, *Stochastics* 22 (1987), 251-288.

Mazziotto, G., and Szpirglas, J., Equations du filtrage pour un processus de Poisson mélangé à deux indices, *Stochastics* 4 (1980), 89-119.

Mazziotto, G., and Szpirglas, J., Arrêt optimal sur le plan, *Z. Wahrsch. verw. Gebiete*, 62 (1983), 215-233.

Mazziotto, G., Merzbach, E., and Szpirglas, J., Discontinuités des processus croissants et martingales à variation intégrable, *Colloque ENST-CNET, Lect. Notes in Math.* 863 (1981), 59-83.

McKean, H. P., Brownian motion with a several-dimensional time parameter, *Th. Probab. Appl.* 8 (1963), 335-354.

McLeish, D. L., An extended martingale invariance principle, *Ann. Probab.* 6 (1978), 144-150.

Merzbach, E., Stopping for two-dimensional stochastic processes, *Stoch. Proc. and their Appl.* 10, (1980), 49-63.

Merzbach, E., Chemins croissants optionels et théorème de section, *Ann. Inst. H. Poincaré*, 19 (1983), 223-234.

Merzbach, E., Point processes in the plane, *Acta Applicandae Mathematicae* 12 (1988), 79-101.

Merzbach, E., and Nualart, D., Different kinds of two-parameter martingales, *Israel J. of Math.* 52 (1985), 193-208.

Merzbach, E., and Nualart, D., A characterization of the spatial Poisson process and changing time, *Ann. Probab.* 14 (1986), 1380-1390.

Merzbach, E., and Nualart, D., A martingale approach to point processes in the plane, *Ann. Probab.* 16 (1988), 265-274.

Merzbach, E., and Nualart, D., Markov properties for point processes on the plane, *Ann. Probab.* 18 (1990), 342-358.

Merzbach, E., and Saada, D., Stochastic integration for set-indexed processes, *Israel Journal of Math.* to appear (1999).

REFERENCES

Merzbach, E., and Zakai, M., Predictable and dual predictable projections of two-parameter stochastic processes, *Z. Wahrsch. verw. Gebiete* 53 (1980), 263-269.

Merzbach, E., and Zakai, M., Bimeasures and measures induced by planar stochastic integrators, *J. Multiv. Anal.* 19 (1986), 67-87.

Merzbach, E., and Zakai, M., Stopping a two-parameter weak martingale, *Probab. Th. and Rel. Fields* 76 (1987), 499-507.

Merzbach, E., and Zakai, M., Worthy martingales and integrators, *Stat. Prob. Letters* 16 (1993), 391-395.

Métivier, M., Un théorème de Riesz pour mesures stochastiques multi-indices, *C.R. Acad. Sc. Paris* 281 (1975), 277-280.

Métivier, M., *Semimartingales: a course on stochastic processes*, De Gruyter, Berlin - New York, 1982.

Métivier, M., and Pellaumail, J., On Doleans-Föllmer's measure for quasi-martingales, *Illinois J. of Math.* 19 (1975), 491-504.

Métivier, M., and Pellaumail, J., Mesures stochastiques à valeurs dans les espaces L_0, *Z. Wahrsch. verw. Gebiete* 40 (1977), 101-114.

Métivier, M., and Pellaumail, J., *Stochastic Integration*, Academic Press, 1980.

Meyer, P.A., Théorie élémentaire des processus à deux indices, *Colloque ENST-CNET, Lect. Notes in Math.* 863 (1981), 1-39.

Millet, A., Convergence and regularity of strong martingales, *Colloque ENST-CNET, Lecture Notes in Math.* 863 (1981), 50-58.

Millet, A., On randomized tactics and optimal stopping in the plane, *Ann. Probab.* 13 (1985), 946-965.

Millet, A., and Sucheston, L., Convergences of classes of amarts indexed by directed sets, *Canad. J. Math.* 32 (1980a), 86-125.

Millet, A., and Sucheston, L., On convergence of L^1-bounded martingales indexed by directed sets, *Probab. and Math. Stat.* 1 (1980b), 192-196.

Millet, A., and Sucheston, L., On regularity of multiparameter amarts and martingales, *Z. Wahrsch. verw. Gebiete* 56 (1981), 21-45.

Millet, A., and Sucheston, L., On fixed point and multiparameter ergodic theorems in Banach lattices, *Canad. J. Math.* 40 (1988), 429-458.

Mishura, Y. S., Two-parameter semi-martingales and point random fields, *Th. Probab. Math. Stat.* 23 (1981), 117-126.

Mishura, Y. S., On properties of the quadratic variation of two-parameter strong martingales, *Th. Probab. and Math. Stat.* 25 (1982), 99-106.

Mishura, Y. S., On some properties of discontinuous two-parameter martingales, *Th. Probab. and Math. Stat.* 29 (1984), 87-100.

Mishura, Y. S., A generalized Itô formula for two-parameter martingales, I and II, *Th. Probab. and Math. Stat.* 30 (1985), 127-142 and 32 (1986), 77-94.

Mishura, Y. S., Canonical representation of two-parameter strong semi-martingales, *Th. Probab. and Math. Stat.* 33 (1986), 91-95.

Mishura, Y. S., Sufficient conditions for relative compactness of measures corresponding to two-parameter strong martingales, *Th. Probab. Math. Stat.*, 134 (1987), 117-125.

Mishura, Y. S., Exponential formulas and Doléans equation for discontinuous two-parameter processes, *Th. Probab. Appl.*, 33, No. 2 (1988), 388-392.

Mishura, Y. S., A martingale characterization of diffusion random fields on the plane, *Th. Probab. Appl.*, 35 (1991), 457-485.

Mishura, Y. S., and Gushchin, A. A., Two-parameter strong martingales: Inequalities for quadratic variation and some decompositions, *Probab. Th. Math. Stat.* 2, B. Grigelionis et al. (eds.), (1990), 181-192, VSP/Moklas.

Molchanov, I. S., *Limit theorems for unions of random closed sets*, Lect. Notes in Math. 1561, Springer-Verlag, New York, 1993.

Moran, P. A. P., Another quasi-Poisson plane point process, *Z. Wahrsch. verw. Gebiete* 32 (1976), 269-272.

Moricz, F., Moment inequalities for the maximum of partial sums of random fields, *Acta Sci. Math. Hung.* 39 (1977), 353-366.

Morkvenas, R., Invariance principle for martingales on the plane (Russian), *Litovskiy Mat. Sbornik* 24 (1984), 127-132.

Morkvenas, R., Convergence of two-parameter stochastic processes (Russian), *Litovskiy Mat. Sbornik* 27 (1987), 724-730.

Nair, M. G., Random space change for multi-parameter point processes, *Ann. Probab.* 18 (1990), 1222-1231.

Neuhaus, G., On weak convergence of stochastic processes with multi-dimensional time parameter, *Ann. Math. Stat.* 42 (1971), 1285-1295.

Neveu, J., *Discrete Parameter Martingales*, North-Holland, Amsterdam, 1975.

Neveu, J., Processus ponctuels, *Lect. Notes in Math.* 598 (1977), 250-447, Springer-Verlag, Berlin-Heidelberg-New York.

Norberg, T., A coordinate-free description of multiparameter stochastic integration, preprint *The University of Göteberg*, 1988.

Norberg, T., Existence theorems for measures on continuous posets, with applications to random set theory, *Math. Scand.* 64 (1989a), 15-51.

Norberg, T., Stochastic integration on lattices, *Tech. Rep. Chalmers Univ. Tech. and The University of Göteberg* (1989b).

Nosko, V. P., Local structure of Gaussian random fields in the vicinity of high level light sources, *Sov. Math. Dokl.* 10 (1969), 1481-1484.

Nualart, D., Decomposition of two parameter martingales, *Stochastics* 5 (1981a), 133-150.

Nualart, D., Weak convergence to the law of two-parameter continuous processes, *Z. Wahrsch. verw. Gebiete* 55 (1981b), 255-259.

Nualart, D., On the quadratic variation of two-parameter continuous martingales, *Ann. Probab.* 12 (1984), 445-457.

Nualart, D., Randomized stopping points and optimal stopping on the plane, *Ann. Probab.* 20 (1992), 883-900.

Nualart, D., and Sanz, M., The conditional independence property in filtrations associated to stopping lines, *Colloque ENST-CNET, Lect. Notes in Math.* 863 (1981a), 202-210.

Nualart, D., and Sanz, M., Changing time for two-parameter strong martingales, *Ann. Inst. H. Poincaré* 17 (1981b), 147-163.

Nualart, D., and Sanz, M., Malliavin calculus for two-parameter processes, *Ann. Sci. Univ. Clermont-Ferrand II* 85 (1985), 73-86.

Nualart, D., and Utzet, F., A property of two-parameter martingales with path-independent variation, *Stoch. Proc. and their Appl.* 24 (1987), 31-49.

Nualart, D., and Yeh, J., Existence and uniqueness of a strong solution to stochastic differential equations in the plane with stochastic boundary processes, *J. Multiv. Anal.* 28 (1989), 149-179.

Nualart, D., and Zakai, M., Multiple Wiener-Itô integrals possessing a continuous extension, *Probab. Th. Rel. Fields* 85 (1990), 131-145.

Nualart, D., Sanz, M., and Zakai, M., On the relations between increasing functions associated with two-parameter continuous martingales, *Stoch. Proc. and their Appl.*, 34 (1990), 99-119.

Orey, S., and Pruitt, W., Sample functions of the N-parameter Wiener process, *Ann. Probab.* 1 (1973), 138-163.

Orsingher, E., On the maximum of random fields represented by stochastic integrals over circles, *J. Appl. Probab.* 24 (1987), 574-585.

Ossiander, M., A central limit theorem under metric entropy with L^2 bracketing, *Ann. Probab.* 15 (1987), 897-919.

Ossiander, M., and Pyke, R., Lévy's Brownian motion as a set-indexed process and related central limit theorems, *Tech. Rep.* 42 (1984), University of Washington, Seattle, Washington.

Papangelou, F., The conditional intensity of general point processes and an application to line processes, *Z. Wahrsch. verw. Gebiete* 28 (1974), 207-226.

Papangelou, F., Point processes on spaces of flats and other homogeneous spaces, *Math. Proc. Cambridge Philos. Soc.* 80 (1976), 297-314.

Paranjape, S. R., and Park, C., Distribution of the supremum of the two-parameter Yeh-Wiener process on the boundary, *J. Appl. Probab.* 10 (1973), 875-880.

Park, W. J., A multi-parameter Gaussian process, *Ann. Math. Stat.* 41 (1970), 1582-1595.

Park, W. J., On Strassen's version of the law of the iterated logarithm for the two-parameter Gaussian process, *J. Multiv. Anal.* 4 (1974), 479-485.

Pitt, L. D., A Markov property for Gaussian processes with a multi-dimensional parameter, *Arch. Rat. Mech. and Anal.* 43 (1971), 367-391.

Pomarede, J. M. L., A unified approach via graphs to Skorokhod's topologies on the function space D, Ph.D. thesis, Yale University (1976).

Prum, B., Properties of Hida processes on R^2, II: Prediction and interpolation problems for processes on R^2, *J. Multiv. Anal.* 15 (1984), 361-382.

Pyke, R., *Multidimensional empirical processes: Some comments on statistical inference and related topics*, M. L. Puri (ed.), Academic Press (1975), 45-48.

Pyke, R., A uniform central limit theorem for partial sum processes indexed by sets, *London Math. Soc. Lect. Notes* 79 (1983), 219-240.

Pyke, R., Opportunities for set-indexed empirical and quantile processes in inference, *Bull. Int. Stat. Inst.* 51 (1985), 1-11.

Rao, K. M., Quasimartingales, *Math. Scand.* 24 (1969), 79-92.

Rāutu, G., A Feller semigroup approach of planar Poisson random measure, *Stud. Cerc. Math.* 39 (1987), 60-70.

Roitgarts, A.D., Some properties of random point fields with nonrandom compensators, *Th. Probab. and Math. Stat.* 36 (1988), 133-142.

Rosen, J., Self intersections of random fields, *Ann. Probab.* 12 (1984), 108-119.

Rosen, J., Continuity and singularity of the intersection local time of stable processes in R^2, *Ann. Probab.* 16 (1988), 75-79.

Rozanov, Y. A., On Gaussian fields with given conditional distributions, *Theory Probab. Applic.* 12 (1967), 381-391.

Rozanov, Y. A., *Markov Random Fields,* Springer-Verlag, Berlin, 1982.

Rozanov, Y. A., Some boundary value problems for generalized random fields, *Theory Probab. Applic.* 35 (1990), 707-724.

Rozanski, R., Markov stopping sets and stochastic integrals. Application in sequential estimation for a random diffusion field, *Stoch. Proc. and their Appl.* 32 (1989), 237-252.

Saavedra, E., C-tightness criterion for non-adapted random fields, *Stoch. Proc. and their Appl.* 46 (1993), 213-218.

Sanz-Solé, M., Local time for two-parameter continuous martingales with respect to the quadratic variation, *Ann. Probab.* 16 (1988), 778-792.

Sanz-Solé, M., R-variations for two-parameter continuous martingales and Itô's formula, *Stoch Proc. and their Appl.* 32 (1989), 69-92.

Shieh, R.N., Results on square functions and quadratic variations of multiparameter martingales, *Math. Rep. Tayoma Univ.* 5 (1982), 85-94.

Skorokhod, A. V., Limit theorems for stochastic processes, *Th. Probab.*

Appl. 1 (1956), 261-289.

Slonowsky, D., Central Limit Theorems for Set-Indexed Strong Martingales, Ph.D. Thesis, University of Ottawa, 1998.

Slonowsky, D., Set-indexed martingales: decompositions and quadratic variation, *Tech. Rep. 332, Lab. Research in Statistics and Probab.*, Carleton Univ. and Univ. of Ottawa (1999).

Slonowsky, D., and Ivanoff, B. G., A central limit theorem for set-indexed strong martingales, *Tech. Rep. 332, Lab. Research in Statistics and Probab.*, Carleton Univ. and Univ. of Ottawa (1999).

Smythe, R. T., Sums of independent random variables on partially ordered sets, *Ann. Probab.* 2 (1974), 906-917.

Smythe, R.T., Multiparameter sub-additive processes, *Ann. Probab.* 4 (1976), 772-782.

Stoica, L., On two-parameter semimartingales, *Z. Wahrsch. verw. Gebiete* 45 (1978), 257-268.

Stoyan, D., Kendall, W. S., and Mecke J., *Stochastic Geometry*, Wiley, Chichester and Akademie-Verlag, Berlin, 1987.

Straf, N. L., Weak convergence of stochastic processes with several parameters, *Proc. Sixth Berkeley Symp. Math. Stat. Probab.* 2 (1972), 187-221.

Strait, R. T., On Berman's version of the Lévy-Baxter theorem, *Proc. A.M.S.* 23 (1966), 91-93.

Stricker, C., and Yor, M., Calcul stochastique dépendant d'un paramètre, *Z. Wahrsch. verw. Gebiete* 45 (1978), 109-133.

Sucheston, L., On one-parameter proofs of multi-parameter convergence theorems, *Z. Wahrsch. verw. Gebiete* 63 (1983), 43-49.

Sznitman, A. S., Martingales dépendant d'un paramètre: une formule d'Itô, *Z. Wahrsch. verw. Gebiete* 60 (1982), 41-70.

Tanaka, Two-parameter optimal stopping problem with switching cost, *Stoch. Proc. and their Appl.*, 36 (1990), 153-164.

Tjostheim, D., Statistical spatial series modelling, *Adv. in Applied Probab.* 10 (1978), 130-154.

Tjostheim, D., Statistical spatial series modelling II. Some further results on unilateral lattice process, *Adv. in Applied Probab.* 15 (1983), 562-584.

Tudor, C., Remarks on the martingale problem in the two-dimensional time parameter, *Revue Roumaine Math. Proc. et Appliquées* 25 (1980), 1551-1556.

Utzet, F., On a nonsymmetric operation for two-parameter martingales, *Ann Sci. Univ.*, Clermont-Ferrand 4 (1985), 113-130.

Valadier, M., Multi-applications mesurables à valeurs convexes compactes, *J. Math. Pures Appl.* 50 (1971), 265-297.

Van der Hoeven, P. C. T., *On Point Processes*, Mathematical Center Tracts, 165, Amsterdam, 1983.

Vanderbei, R. J., Towards a stochastic calculus for several Markov processes, *Adv. in Appl. Math.* 4 (1983), 125-144.

Vares, M., Representation of the square integrable martingales generated by a two-parameter Lévy process, *Stochastics* 15 (1985), 311-333.

Walsh, J. B., The local time of the Brownian sheet, *Astérisque* 52-53 (1978), 47-61.

Walsh, J. B., Convergence and regularity of multiparameter strong martingales, *Z. Wahrsch. verw. Gebiete* 46 (1979), 177-192.

Walsh, J. B., Optional increasing paths, *Colloque ENST-CNET, Lect. Notes in Math.* 863 (1981a), 172-201.

Walsh, J. B., A stochastic model of neural response, *Adv. in Appl. Probab.* 13 (1981b), 231-281.

Walsh, J. B., Propagation of singularities in the Brownian sheet, *Ann. Probab.* 10 (1982), 279-288.

Walsh, J. B., An introduction to stochastic partial differential equations, *Lect. Notes in Math.* 1180 (1986a), 266-439.

Walsh, J. B., Martingales with a multi-dimensional parameter and stochastic integrals in the plane, *Lect. Notes in Math.* 1215 (1986b), 329-491.

Wang, Z., and Xue, X., On convergence of vector-valued mils indexed by a directed set. In: *Almost Everywhere Convergence (Proceedings)*, G. A. Edgar and L. Sucheston (eds.), Academic Press, 1989, 405-416.

Washburn, R. B., Jr., and Willsky, A. S., Optional sampling of submartingales indexed by partially ordered sets, *Ann. Probab.* 9 (1981), 957-970.

Wichura, M. J., Inequalities with applications to the weak convergence of random processes with multi-dimensional time parameters, *Ann. Math. Stat.* 40 (1969), 681-687.

Wichura, M. J., Some Strassen-type laws of the iterated logarithms for multiparameter stochastic processes with independent increments, *Ann. Probab.* 1 (1973), 272-296.

Wong, E., Homogeneous Gauss-Markov random fields, *Ann. Math. Stat.* 40 (1969), 1625-1635.

Wong, E., Recent progress in stochastic processes: A survey, *Trans. on Inf. Th.* 19 (1973), 262-274.

Wong, E., A likelihood ratio formula for two-dimensional random fields, *IEEE Trans. Inf. Theory* 20 (1976), 418-422.

Wong, E., Recursive causal filtering for two-dimensional random fields, *IEEE Trans. Inf. Th.* 24 (1978), 50-59.

Wong, E., and Zakai, M., Martingales and stochastic integrals for processes with a multidimensional parameter, *Z. Wahrsch. verw. Gebiete* 29 (1974), 109-122.

Wong, E., and Zakai, M., Weak martingales and stochastic integrals in the plane, *Ann. Probab.* 4 (1976), 570-586.

Wong, E., and Zakai, M., Likelihood ratios and transformation of probability associated with two-parameter Wiener processes, *Z. Wahrsch. verw. Gebiete* 40 (1977), 238-308.

Wong, E., and Zakai, M., Markov processes on the plane, *Stochastics* 15 (1985), 311-333.

Wong, E., and Zakai, M., Multiparameter martingale differential forms, *Probab. Th. and Rel. Fields* 74 (1987), 429-453.

Wschebor, M., *Surfaces Aléatoires*, Lect. Notes in Math. 1147, Springer-Verlag, New York, 1985.

Yaglom, A. M., Some classes of random fields in n-dimensional space related to stationary random processes, *Th. Probab. and its Appl.* 2 (1957), 273-320.

Yeh, J., Stopping times and an extension of stochastic integrals in the plane, *J. Multiv. Anal.* 11 (1981), 334-345.

Yeh, J., Two parameter stochastic differential equations, in *Real and Stochastic Analysis*, M. M. Rao (ed.), Wiley-Interscience (1986), 249-344.

Yor, M., Representation de martingale de carré intégrable relatives aux processus de Wiener et de Poisson à n paramètres, *Z. Wahrsch. verw. Gebiete* 35 (1976), 121-129.

Yor, M., Sur la représentation comme intégrales stochastiques des temps d'occupation du mouvement brownien dans R^d, *Sém. de Probab. XX*, Lect. Notes in Math. 1204 (1986), 543-552.

Yoshida, Y., On an optimal stopping problem for multi-parameter diffusion processes, *J. of Info. and Optimiz. Sciences,* 11 (1990), 473-492.

Zakai, M., Some classes of two-parameter martingales, *Ann. Probab.* 9 (1981), 255-265.

Zayats, V., A CLT in a Hilbert space and correlation function estimates over the whole parameter set, in: *Limit Theorems in Probability and Statistics,* I. Berkes, E. Csaki and P. Revesz (eds.), North Holland, 1990, 545-556.

Zbăganu, G., A generalization of F4-hypothesis of Cairoli and Walsh, *Rev. Roumaine Math. Pure Appl.* 31 (1986), 919-922.

Zbaganu, G., and Zhuang, X.W., Two-parameter filtrations with respect to which all martingales are strong, *Z. Wahrsch. verw. Gebiete* 61 (1982), 437-452.

Zhou, J.W., The strong Markov property of two-parameter processes, *Chinese J. Appl. Probab. Stat.* 2 (1986), 302-306.

Zimmerman, G.J., Some sample function properties of the two-parameter Gaussian process, *Ann. Math. Stat.* 43 (1972), 1235-1246.

Index

$B(\mathcal{A})$, 134
$C(\mathcal{A})$, 134
$D([0,\infty))$, 133
$D[0,1]$, 17
D_H, 138
d, 9
d_G, 138
d_H, 20, 134
$F'_Q(0)$, 171
$G(x)$, 138
g_n, 10
$J(y)$, 176
L^p, 25, 27, 56
$M_f(X)$, 149
$r(\mathcal{P}_0^*)$, 36
$r(\mathcal{P}_0)$, 36, 38
$S(\mathcal{A})$, 148
T, 9
t_A, 48
$\emptyset'$, 11
ϵ-twins, 154
μ_X, 75, 79, 80
$\mu_X^{(2)}$, 76, 81
Π, 46
$\sum_X(\mathcal{E})$, 114
τ^f, 63
$\tilde{\mathcal{A}}(u)$, 11
$\Xi(\Gamma, \Sigma, \Delta)$, 139
$\mathcal{A}$, 9
$\mathcal{A}(u)$, 9
$\mathcal{A}_n$, 10
$\mathcal{A}_n(u)$, 10
$\mathcal{B}$, 9
$\mathcal{C}^\ell(\mathcal{A}')$, 14
$\mathcal{C}$, 12

$\mathcal{C}(u)$, 13
$\mathcal{C}_n$, 13
$\mathcal{D}(\mathcal{A})$, 133, 134
$\mathcal{D}[S(\mathcal{A})]$, 148
$\mathcal{D}^1(\mathcal{A})$, 153
$\mathcal{F}_A$, 22
$\mathcal{F}_B^0$, 22
$\mathcal{F}_B^r$, 22
$\mathcal{F}_\xi$, 31
$\mathcal{F}_{\xi^-}$, 31
$\mathcal{G}$, 138
$\mathcal{G}_C^*$, 23
$\mathcal{G}_C$, 23
$\mathcal{K}$, 9
$\mathcal{O}$, 9
$\mathcal{P}^*$, 35
$\mathcal{P}$, 35
$\mathcal{P}(\mathcal{A})$, 136
$\mathcal{P}_0^*$, 35
$\mathcal{P}_0$, 35
$\pi_{A_1,\cdots,A_n}$, 145

adapted process
    T-indexed, 48, 82
    $\mathcal{A}$-indexed, 25
additive
    function, 25
    process, 25
admissible
    function, 75
    measure, 79, 80
    square function, 76
    square measure, 81
Aldous' stopping time condition, 160

announcable stopping set, 35, 40
announcing sequence, 40
asymptotic rarefaction of jumps, 175
atoms, 136
  masses, 136

big bang, 15
Borel sets, 9
bounded sequence in $\mathcal{A}$, 10
Brownian motion, 66, 67
  martingale characterization, 106
  quadratic variation, 86
  variance measure, 67

central limit theorem, 175
  functional, 180
  semi-functional, 176
CI, 23
class $(D')^*$, 78
class D, 63, 78
class D', 78
compactness in $\mathcal{D}(\mathcal{A})$, 142
compensator, 73
  *-compensator, 73
  *-predictable, 82, 85
  discrete approximation, 86
  predictable, 82, 85
conditional independence
  CI property, 23
  F4 property, 23
  of σ-algebras, 23
consistent with boundaries, 140
consistent with the strong past, 14
continuity set, 140, 153
  stochastic, 158
convergence
  functional, 146
    on flows, 133
  in finite dimensional distribution, 133, 146
  semi-functional, 133, 149
  vague, 153
  weak, 144
    in $\mathcal{D}(\mathcal{A})$, 146
Cox process, 70
  compensator, 92
  limit theorem, 168

derivative from quadrant, 171
determining class, 145
dilation, 41
discrete
  process, 87
  stopping set, 29
dissecting system, 14
distributive lattice, 15
domain, 43
Doob-Meyer decomposition, 73, 81, 85, 116
dual predictable projection, 117
début, 39

empirical process, 169
  Poisson limit theorem, 170
  weighted, 181
    CLT, 184
extremal representation, 13

F4, 23
filtration, 22
  minimal, 66
  minimal outer-continuous, 26
finite dimensional projections, 145
fixed atoms, 70
flow, 98
  continuous, 98
  right-continuous, 98
  simple, 148
function
  additive, 25
  admissible, 75
  admissible square, 76
  increasing, 26

INDEX 209

monotone inner-continuous, 101
monotone outer-continuous, 26
monotone outer-continuous on $\mathcal{A}$, 26
outer continuous with inner limits, 134
purely atomic, 136
functional central limit theorem, 180
functional convergence, 146
  on flows, 133
functional limit theorem, 146

graph function, 145

Hausdorff
  distance, 19
  metric, 19, 20, 134
  topology, 19
history
  strong, 23
  weak, 23

incomplete closed graph, 138
increasing
  $\mathcal{A}_n$-increasing process, 87
  function, 26
  process, 27
independent increments, 66
indexing collection, 10
indistinguishable processes, 57
inner limits, 134
integrable process, 25

jump functional, 176
jump point, 152

Krickeberg decomposition, 59

lattice separability, 48
left-continuous
  function, 47, 48
  process, 47, 49
left-neighbourhood, 14
Lévy characterization, 97
local
  martingale, 114
  strong martingale, 114
  submartingale, 114
  weak martingale, 114
localization, 113
localizing sequence, 114
locally of class D, 114
lower set, 15, 17
  convex, 17

martingale, 53, 54
  $\mathcal{A}_n$-martingale, 87
  characterization, 97
  local, 114
  strong, 54
  strong submartingale, 54
  strong supermartingale, 54
  submartingale, 54
  supermartingale, 54
  weak, 54
  weak submartingale, 54
  weak supermartingale, 54
maximal representation, 13
metric entropy, 133
minimal, 14
minimal filtration, 66
  outer-continuous, 26
monotone inner-continuous function, 101
monotone outer-continuous
  L^p-monotone outer-continuous process, 27
  L^p-monotone outer-continuous process on $\mathcal{A}$, 27
  filtration, 22
  function, 26
  function, on $\mathcal{A}$, 26
  process, 27
  process, on $\mathcal{A}$, 27

numbering consistent with the strong past, 14

optional sampling theorem, *see stopping theorem*, 61
outer-continuous, 134

point measure, 152
   simple, 153
   strictly simple, 154
point process, 91
   simple, 91, 158
   strictly simple, 158
Poisson limit theorem
   fidi, 161
   functional, 168
Poisson process, 66, 70, 160
   compensator, 92
   doubly stochastic, *see Cox process*, 70
   martingale characterization, 107
   mean measure, 70
potential, 58
predictable
   σ-algebra, 35
   *-compensator, 82, 85
   *-predictable σ-algebra, 35
   *-predictable increasing process, 82
   *-predictable rectangles, 35
   *-quadratic variation, 86
   compensator, 82, 85
   increasing process, 82
   quadratic variation, 86
   rectangles, 35
process
   L^p, 25
   L^p-bounded, 56
   L^p-monotone outer-continuous, 27
   L^p-monotone outer-continuous on $\mathcal{A}$, 27
   $\mathcal{A}_n$-increasing, 87
   σ-additive, 28
   $\mathcal{A} - L^p$ bounded, 56
   $\mathcal{A}$-indexed, 25
   adapted, 25
   adapted, T-indexed, 48
   additive, 25
   bounded variation, 158
   discrete, 87
   increasing, 27
   integrable, 25
   left-continuous, 47, 49
   monotone outer-continuous, 27
   monotone outer-continuous on $\mathcal{A}$, 27
   point process *see point process*, 91
   purely atomic, 133
   single jump *see single jump process*, 94
   square integrable, 25
progressive σ-algebra, 43, 46
proper sets, 136
purely atomic
   function, 136
      variation, 136
   process, 133

quadrant, 16
   derivative, 171
quadratic variation, 73, 74, 81
   *-predictable, 86
   *-quadratic variation, 74, 86
   discrete approximation, 86
   predictable, 86
quasimartingale, 119
   strong, 119
   weak, 119
quasipotential, 127

random set, 43
   simple, 43
reduce, 114

INDEX

Riesz decomposition, 58, 127

semi-algebra, 12, 36
semi-functional central limit
    theorem, 176
semi-functional convergence, 133,
    149
semilattice, 11
separability from above, 10
set-interval, 137
SHAPE, 13
shatter, 154
sigma-algebra
    *-predictable, 35
    weak predictable, 35
    events prior to a stopping set,
        31
    events strictly prior to a
        stopping set, 31
    predictable, 35
    progressive, 43, 46
simple stopping set, 28
single jump process, 94
    *-compensator, 95
Skorokhod
    J_1 topology, 133
    J_1-topology
        tightness, 160
    J_2 topology, 133
    function space, 133
square integrable process, 25
stochastic
    continuity set, 158
    interval, 36
        closed, 40
        inner-open and outer-closed,
            37
        open, 40
        outer-open and inner-closed,
            40
    process, see process, 25
stopping
    line, 43
    point, 43

set, see stopping set, 28
theorem, see stopping theorem,
    61
time, 28
stopping set, 28
    σ-algebra of events prior to, 31
    σ-algebra of events strictly
        prior to, 31
    announcable, 35, 40
    discrete, 29, 37
    simple, 28
stopping theorem, 61
    characterization of martingale,
        63
    characterization of strong
        martingale, 61
    for weak martingale, 65
strictly from above, 10
strictly simple, 151
    point measure, 154
    point process, 158
strong
    $\mathcal{A}_n$-martingale, 87
    martingale, 54
        characterization, 61
        local, 114
    quasimartingale, 119
    submartingale, 54
    supermartingale, 54
submartingale, 54
    $\mathcal{A}_n$-submartingale, 87
    local, 114
supermartingale, 54

tightness, 133
    on $D[0, 1]$, 160
    on $\mathcal{D}(\mathcal{A})$, 146
TIP, 48

vague
    convergence, 153
    topology, 151, 153
Vapnik-Červonenkis class, 68

Watanabe characterization, 97
way below, 48
weak
   $\mathcal{A}_n$-martingale, 87
   martingale, 54
      local, 114
   quasimartingale, 119
   submartingale, 54
   supermartingale, 54
weak convergence in $\mathcal{D}(\mathcal{A})$, 144, 146
weak predictable σ-algebra, 35
weighted empirical process, 181
   CLT, 184